PRAISE FOR *CELLS ARE THE NEW CURE*

"*Cells Are the New Cure* explains the world of cutting-edge complex scientific discovery in easy-to-understand language, interweaving the human element through stories of remission and cure. As someone who is blessed to be thriving since my stem cell transplant, this book gives us hope."

—ROBIN ROBERTS, *GOOD MORNING AMERICA*

"*Cells Are the New Cure* details the recent discoveries of the world's top scientists and clinicians in the rapidly advancing field of cellular medicine. This paradigm shift in medicine will change the way we treat disease in the very near future and advances in immunotherapy will revolutionize how we approach cancer."

—SEAN PARKER, PRESIDENT, PARKER FOUNDATION

"Robin Smith and Max Gomez have illuminated one of the great frontiers in medicine for the general reader: The way aberrant cellular biology leads to disease and how cellular therapy can be harnessed to treat it. *Cells Are the New Cure* describes the many novel treatments and therapies under development and in use in the clinic and catalogues the potential of cellular therapy to alleviate untold pain and suffering across many lethal diseases like cancer and age-related degenerative conditions. They describe these historic scientific advances while compassionately relaying the real-life experiences of patients who have benefited from them."

—RONALD A. DEPINHO, MD, FORMER PRESIDENT, UT MD ANDERSON CANCER CENTER

"*Cells Are the New Cure* explores the potential in modern medicine to harness the human body's own curative abilities to battle cancer, autoimmune diseases, and many rare diseases. Set against the backdrop of real-life stories of patients who have benefited from these therapies while they are undergoing clinical development, the book highlights that unique moment where science and life intersect."

—STEPHEN C. GROFT, PHARMD., SENIOR ADVISOR TO THE DIRECTOR, NATIONAL CENTER FOR ADVANCING TRANSLATIONAL SCIENCES, AND FORMER DIRECTOR, OFFICE OF RARE DISEASES RESEARCH, NATIONAL INSTITUTES OF HEALTH

"*Cells Are the New Cure* writes a prescription for hope, long missing from the medical care of patients like me. I have known too many dangerous conditions in my adult life. Now, science is stepping in with cutting-edge research and clinical trials that offer a bright future for those of us who have been denied a view of life beyond today."

—RICHARD M. COHEN, JOURNALIST AND BESTSELLING AUTHOR WHO HAS BATTLED ILLNESSES FOR MORE THAN FORTY YEARS

CELLS ARE THE NEW CURE

CELLS ARE THE NEW CURE

The Cutting-Edge Medical Breakthroughs That Are Transforming Our Health

ROBIN L. SMITH, MD + MAX GOMEZ, PHD

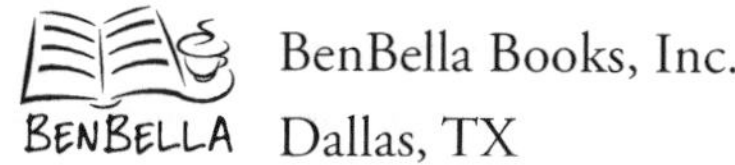

BenBella Books, Inc.
Dallas, TX

BenBella Books, Inc.
10440 N. Central Expressway, Suite 800
Dallas, TX 75231
www.benbellabooks.com
Send feedback to feedback@benbellabooks.com

BenBella is a federally registered trademark.

Printed in the United States of America

ISBN 978-1-63774-582-3 (trade paperback)

The Library of Congress Cataloging-has cataloged the hardcover edition as follows:
Names: Smith, Robin L., 1964- author. | Gomez, Max, author.
Title: Cells are the new cure : the cutting-edge medical breakthroughs that are transforming our health / Robin L. Smith and Max Gomez.
Description: Dallas, TX : BenBella Books, Inc., [2017] | Includes bibliographical references and index.
Identifiers: LCCN 2017013255 (print) | LCCN 2017014644 (ebook) | ISBN 9781944648848 (electronic) | ISBN 9781944648800 (trade cloth : alk. paper)
Subjects: | MESH: Cell- and Tissue-Based Therapy | Immunotherapy | DNA Repair | Popular Works
Classification: LCC RM370 (ebook) | LCC RM370 (print) | NLM WO 75 | DDC 615.3/7—dc23
LC record available at https://lccn.loc.gov/2017013255

Editing by Glenn Yeffeth and David Bessmer
Copyediting by Miki Alexandra Caputo
Proofreading by Amy Zarkos and Lisa Story
Indexing by WordCo Indexing Services, Inc.
Cover design by Emily Weigel
Text design and composition by Aaron Edmiston

CONTENTS

Foreword xi
Introduction xv

PART ONE: THE PROMISE OF REGENERATIVE MEDICINE

Without a doubt, stem cell research will lead to the dramatic improvement in the human condition and will benefit millions of people.

—ELI BROAD

CHAPTER 1: Repairing Injured and Aging Tissue 2
CHAPTER 2: Growing New Body Parts in the Lab 20
CHAPTER 3: Repairing the Brain 35
CHAPTER 4: Stem Cells and Cancer 54

PART TWO: OUR IMMUNE SYSTEM AS WEAPON AND HEALER

Natural forces within us are the true healers of disease.

—HIPPOCRATES

CHAPTER 5: Teaching the Body to Fight Cancer 66
CHAPTER 6: Stopping Autoimmune Disease 85

CHAPTER 7: Allergies and Food Sensitivities—Peanuts, Celiac, and Beyond 102
CHAPTER 8: The Future of Immunotherapy 120

PART THREE: CHANGING OUR DNA—RARE DISEASES AND DESIGNER HUMANS

If we could make better humans, why shouldn't we?

—JAMES WATSON, PHD

CHAPTER 9: The Gene Repair Tool Kit 138
CHAPTER 10: Repairing DNA in Rare Diseases 149
CHAPTER 11: Putting a Bull's-Eye on Sick Cells 163
CHAPTER 12: Should We Alter DNA? 176

PART FOUR: HUMAN 2.0

There are no such things as incurables. There are only things for which man has not found a cure.

—BERNARD BARUCH

CHAPTER 13: Can We Prevent or Even Reverse Illness and Aging? 188
CHAPTER 14: The Road to 100 Plus 211

PART FIVE:
BIG MONEY, BIG DATA, AND THE FUTURE OF MEDICINE

The world has been very careful to pick very few diseases for eradication, because it's very tough.

—BILL GATES

CHAPTER 15: Big Data Leads to Better Medicine? **224**

CHAPTER 16: Philanthropy Drives Innovation **237**

CHAPTER 17: Clinical Trials in the Era of Cellular Medicine **256**

CHAPTER 18: Cells Will Be the Drugs of the Future **269**

Acknowledgments **282**

Photo Credits **284**

Index **286**

About the Authors **299**

FOREWORD

One of the first stories I covered as a television reporter was the restriction on US federal funding for research on stem cells. It was August 2001, and for the next eight years, many scientists would spend more time trying to find new sources of funding than actually doing research. When the restrictions were relaxed in March 2009, there was a sense of optimism in the stem cell community but also a feeling the United States was nearly a decade behind other countries and now had to make up a lot of lost ground.

To be fair, supporters of the restrictions felt they were acting in defense of human life in protecting embryos. Furthermore, during this time several lessons were learned in the world of stem cell research that might not otherwise have been learned. First, the early data from embryonic stem cell pioneers did not prove particularly promising. Additionally, induced pluripotent stem cells (iPSCs)—adult cells induced to exhibit the same properties as embryonic stem cells—were discovered. Finally, adult stem cells, thought for sixty years to reside only in bone marrow, were found in many more tissues. As we now know, this greatly opened up the range of therapeutic possibilities.

And yet, in the United States we still feel the hangover of that eight-year drought in stem cell research. Patients are understandably dubious of their true value, as too many charlatans have made laughable claims unsupported by scientific data. While the number of clinical trials has increased, the US medical system doesn't yet offer formal clinical training programs to teach stem cell therapies. My colleagues in medical journalism write with greatly tempered optimism that borders on cynicism whenever we do stories on the promise of stem cells. And at this writing, the US Food and Drug Administration has approved only one stem cell therapy product, Hemacord, to restore low blood cell counts. In fact, the FDA is sometimes stymied at how to regulate adult stem cells at all, especially when they are extracted from and then injected back into the same human body. It seems that adult stem cells don't fall neatly into the definition of device or drug, the two categories on which the FDA focuses.

As things stand now, if you have heard of Americans getting stem cell treatments, they are usually wealthy individuals who have run out of options and are willing to try anything, even if it is untested and unproven. That doesn't inspire a lot of confidence, but it is the story I have often heard in the mainstream and scientific media almost since I started my careers in journalism and neurosurgery.

That was the backdrop when I first received a call from the authors of this book, Robin L. Smith, MD, and Max Gomez, PhD. They wanted me to participate in Cellular Horizons, a conference in Vatican City focused on cell-based therapies. That's right, Vatican City. After fifteen years of bearing witness to the constant collisions of science and theology, the impact on federal funding, and several-year cycles of optimism and pessimism, things were coming full circle. I would hear from scientists, ethicists, and Pope Francis himself about their faith in stem cells.

Make no mistake, the Catholic Church continues to oppose research or therapies that involve the destruction of embryos. It does, however, support adult stem cell research and asked scientists from all over the world to present their data. These were reputable scientists and regulators from Germany, China, Japan, India, and Australia. The United States was also well represented with researchers from major hospitals and universities presenting compelling data on a variety of ailments.

Sabrina Strickland, MD, of the Hospital for Special Surgery in New York City is using stem cells as a viable alternative for painful osteoarthritis. Eduardo Marban, MD, PhD, at Cedars-Sinai Heart Institute in Los Angeles used them to reduce scar tissue size in the muscle after a heart attack. One of the more provocative trials I heard described is taking place at Duke, where Joanne Kurtzberg, MD, is looking into the use of stem cells for children with autism. While not all stem cell researchers agree this is a rational application of stem cell therapy, there is an increasing number of children going to profit-driven clinics that offer up no data. The Duke trial will at least produce some answers about the impact of cellular therapy on neurons in the brain at various stages of development.

As a journalist, I was particularly interested in meeting the patients behind all the data. Having seen the enormous challenges of autoimmune disease in my own family, I agreed to moderate a panel with Richard Burt, MD, from Northwestern University and his beautiful young patients Grace Meihaus and Elizabeth Cougentakis. At age seventeen, Grace developed rapidly progressive systemic scleroderma that not only hardened her skin but caused multiple internal organs to become scarred and inflamed. Out of options, she found Dr. Burt and underwent an autologous hematopoietic stem cell transplant. Her symptoms improved within months, and she was able to return to college and her way of life. Elizabeth's myasthenia gravis, another autoimmune disease, first affected the small muscles of her eyes, then progressed to a point where she could no longer feed herself or walk. Within six months, she was on a breathing machine and being fed with a tube. If I hadn't heard her describe it as vividly as she did, I would not have believed that Elizabeth started to feel better immediately after treatment as she regained control of her eyes and other motor functions. Eight years later, she is completely healthy and requires no medications. As a father of three daughters, I was buoyed and encouraged by what the future could hold for them and other children around the world if they ever became ill.

We are a long way from seeing these therapies become widely applied. There are still appropriate scientific, regulatory, and cultural hurdles to overcome. And yet, it has become increasingly clear that someday soon even the most ardent critics will come to support the belief that our greatest healing tools may lie within the human body itself. As this field is still very much in its infancy, it is hard to gather all the knowledge in one place. Books like this should be seen as part of a living, breathing body of knowledge that will grow, evolve, and be reborn. What will never change, however, is the measured yet relentless optimism these pages bring.

—SANJAY GUPTA, MD

MARCH 2017

INTRODUCTION

It is health that is real wealth and not pieces of gold and silver.
—MAHATMA GANDHI

Imagine a world where victims of irreversible spinal cord injuries are able to get out of their hospital beds and walk. Where there's no shortage of donor organs for transplants—because we no longer need donors—we just grow new organs in laboratories. Where stem cells from bone marrow or fat are used to repair damaged hearts and spinal discs and lungs—and even to slow the progression of Lou Gehrig's disease and Parkinson's. Where your immune system can be taught to destroy cancer. Where your DNA can be repaired before symptoms of genetic diseases even start. In short, imagine a world of longer and healthier lives.

This is the promise of cell therapies. What once seemed like science fiction, pipe dreams, or outright miracles is already becoming reality. Cell therapies are now being used to combat cancer, repair injuries, and treat autoimmune conditions. Researchers are making replacement organs with 3D bioprinters where scaffolds are sprayed with layers of living cells. Human trials of cell therapies for Alzheimer's, ALS, and Parkinson's are underway. And scientists are experimenting with cell-based antiaging therapies that will help us not only to live longer but, in effect, to stay young longer.

Leading experts expect that this rapidly developing field will soon transform the treatment of many diseases that have caused degeneration and death throughout human history—with faster, more complete recovery and significantly fewer risks and side effects. This research is ongoing with over 35,000 clinical trials underway across the globe.

A decade ago, some of the advances you will read about in this book seemed like the harebrained ideas of fringe scientists. Today they are saving lives. Others are poised to emerge from the laboratory as accepted treatments. Still others, based on fascinating new understandings of cells and diseases, will take longer to reach widespread use.

We are on the precipice of a revolution in medicine. Cell therapies represent a new frontier, harnessing the power of our own biology as the new "drugs" that will not merely treat but actually cure disease.

PART ONE
THE PROMISE OF REGENERATIVE MEDICINE

Without a doubt, stem cell research will lead to the dramatic improvement in the human condition and will benefit millions of people.

—ELI BROAD

CHAPTER ONE

REPAIRING INJURED AND AGING TISSUE

Research serves to make building stones out of stumbling blocks.

—ARTHUR D. LITTLE

In the Greek myth, Prometheus steals fire from Olympus, hiding it inside a giant fennel stalk and delivering it to mankind. As punishment, Zeus has Prometheus chained to a stake on Mount Kaukasos, where every day a giant eagle eats his liver and every night his liver regrows. The story shares themes with mythologies around the world. Many cultures describe a superhero or trickster who steals fire, and since prehistory humans have fantasized about the ability to regrow or repair our bodies. The origin of this fantasy is obvious—the breakdown and eventual failure of tissue is a central fact of human existence. We age. And as we age, our bodies progressively break down until eventually we die. Functions that are lost to time, disease, or injury rarely return.

But now we are tantalizingly close to rewriting the narrative describing the decline of human tissue. The myth of tissue regeneration is becoming reality. The secret is proving to be not a new compound or surgical technique, but the power that lies within *our own* human cells. Specifically, scientists and doctors are learning to restart programs of growth and repair buried deep in the cells that built our bodies in the first place. In the womb, stem cells drove the growth of the tissues that became your mind and body. We are now starting to use stem cells for a similar purpose in the adult body—to repair and regenerate these same tissues. Today, innovative scientists are showing that stem cells can indeed repair your body after damage or disease.

Of course, no matter your politics, religion, or beliefs, you have heard about the moral and ethical challenges of embryonic stem cell research and treatments. These are the cells most active in transforming a fertilized egg into a viable human body, and so some research and treatments have depended on stem cells harvested from fetuses. However, recent work is allowing doctors and patients to sidestep this ethical dilemma. Increasingly, treatments are based on stem cells derived

CELL DIVISION

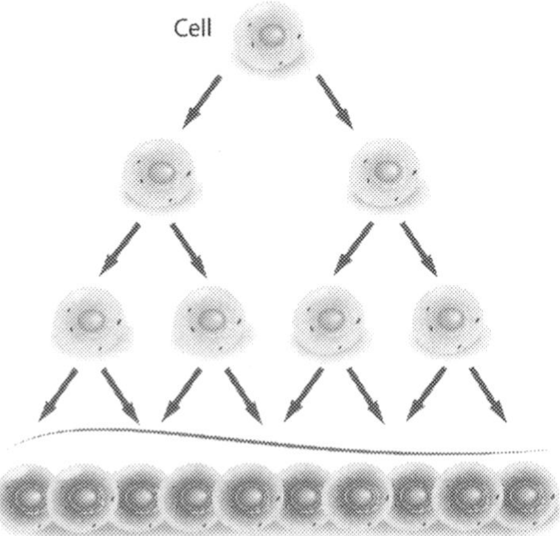

The normal life cycle of a cell proceeds through fifty to seventy cell divisions until it reaches "old age" or senescence, stops dividing, and eventually dies off. Scientists are developing ways to slow, stop, and even reverse this process to keep cells alive and healthy beyond their natural life span.

from *adult* tissues, including your own skin, fat, or blood cells. Also, promising work seeks to strip adult cells of their "differentiation" to return them to a more stem-like, even embryonic-like, state, so that they can be used to regenerate various kinds of tissues.

This new field of regenerative medicine borrows from biology, biochemistry, chemistry, tissue engineering, physics, and applied engineering—an interdisciplinary approach to stimulating the body to heal and repair itself. In addition to rejuvenating a creaky knee or bad back, regenerative medicine has the potential to create therapies for previously untreatable diseases and conditions.

Think of the hundreds or even thousands of diseases that result from malfunctioning, damaged, or failing tissues. Each of these may be treatable and even curable through regenerative medicine therapies. Yes, this sounds like an extreme claim, leading some to downplay the potential of regenerative medicine as if it were a nineteenth-century miracle tonic promising to cure everything from dropsy to ague.

However, as evidence mounts for the benefits of cell-based regenerative medicine, what once was fringe science is becoming a recommended part of mainstream treatment for many conditions. The more we learn, the more this list of conditions for which stem cells can repair tissue and restore function grows. But before we get to real evidence regarding conditions that benefit from regenerative medicine, it's important for you to know a little bit about these stem cells that are the key to these new therapies.

THE PROMISE OF STEM CELLS

Life begins for each of us with the fusion of a sperm and an egg. This single cell, called a zygote, is the primal stem cell, the genesis of all the

CELL DIVISION

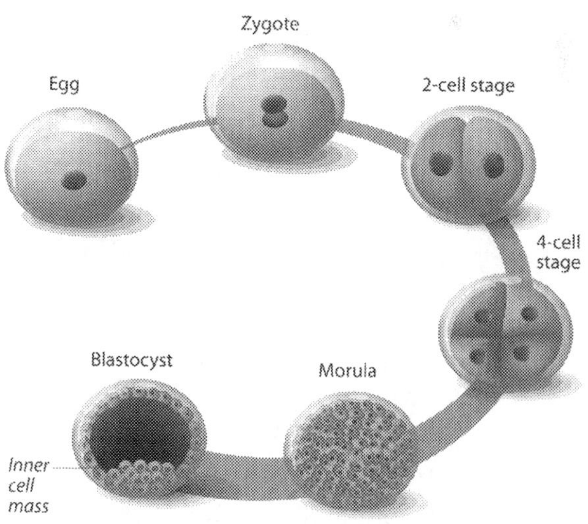

Once an egg is fertilized, it becomes a zygote and begins to divide. After three to four days the ball of cells resembles a mulberry, hence morula (Latin for mulberry). A day or two later the cells reorganize into a hollow ball, forming an inner cell mass that is destined to become the actual embryo, and the other cells will become the placenta.

tissues that will become your body. Not long after fertilization, the zygote divides, making two, then four, then eight, then sixteen cells. Once these cells have divided exactly seven times to make 128 cells, the structure is called a blastocyst. At that point, a switch is flipped, and instead of continuing to manufacture duplicates, cells start the process of differentiation. They take a step toward specialization, becoming cells of the ectoderm, mesoderm, and endoderm. These three cell types, which form what are called *germ layers*, will go on to diversify into trillions of cells representing 200 types. The ectoderm gives rise to cells of the nervous system, among other things. The mesoderm forms muscles, bones, and organs. And the endoderm forms cells that line the insides of many of the body's tube-like systems.

Eventually, cells differentiate until some become skin, others become cells that allow the heart muscle to beat in rhythm, and still others become cells in your joints that cushion the bones. Each fully differentiated cell has a specific job to do. These cells can only reproduce copies of themselves, with a few exceptions, which we will discuss. By the time one is born, the vast majority of one's cells have become specialists, but earlier undifferentiated cells never completely vanish. Inside each one of us are cells with various levels of cell-making capabilities, able to make a few or even many types of tissues. These are stem cells.

Though stem cells exist across a spectrum of differentiation, science and policy generally divide them into two categories: embryonic stem cells and adult stem cells. Embryonic stem cells are the original stem cells. In the lab, they are commonly derived from 128-cell undifferentiated blastocysts created by in vitro fertilization. While their ability to give rise to many tissue types makes them attractive, this same ability leads to a major downside in practice. Unbound by the checks and balances of the body's original development, these pluripotent embryonic stem cells, when placed in areas that are outside their usual developmental context, are prone to causing cancer. When researchers introduce embryonic stem cells into animal models, these animals tend to develop tumors.

However, we can count difficulty of use as only the second challenge in working with embryonic stem cells. "Embryonic stem cells are not going to make it as therapeutic agents, mainly due to ethical reasons," says Eckhard Alt, MD, PhD, the Todd and Linda Broin Distinguished Professor and Chair at Sanford Health in Sioux Falls, South Dakota. Alt directs the Sanford Project, a 20-member type 1 diabetes research team focusing on leveraging stem cells to generate new insulin-producing beta cells in the pancreas.

One solution to the ethical and moral problems of embryonic stem cells is the creation of what are called iPSCs, otherwise known as induced pluripotent stem cells. By stripping adult stem cells of their differentiation, researchers are increasingly able to return them to a state similar to embryonic stem cells. It turns out that by stressing adult stem cells and then exposing them to certain chemicals, they can be coaxed back into a stem-like state, able to give rise not only to clones of their own tissue type but also to a range of other tissues. The discovery of this technique earned its pioneers, Sir John Gurdon, DPhil, DSc, FRS, and Shinya Yamanaka, MD, PhD, the 2012 Nobel Prize in Physiology or Medicine.

However, Alt points out that for many conditions, treatment may not require a cell's ability to create *any* cell type. We might not need true pluripotent cells at all. Instead, more specialized, more differentiated adult stem cells might be used directly to regrow or repair damage in the tissues they naturally support. And since these cells are found in sufficient amounts in adult fat, bone marrow, and other tissues, many patients can be treated with their own cells, avoiding the need for powerful antirejection medications.

These adult stem cells have started down the path of differentiation but have not yet reached the final cell type. Adult stem cells produce the cells that maintain the body's tissues and organs, and, importantly, they can divide and reproduce indefinitely. These cells

Nobel Prize Laureate Sir John Gurdon at the Second International Conference on Regenerative Medicine—A Fundamental Shift in Science & Culture (Vatican, April 2013).

SOURCES OF MESENCHYMAL STEM CELLS

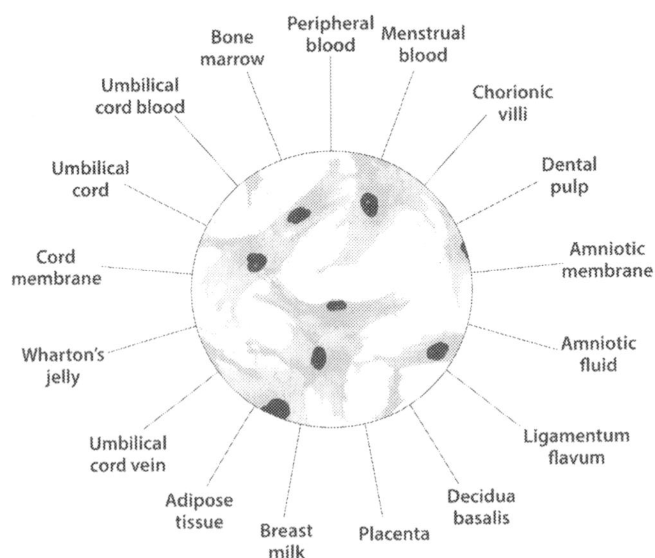

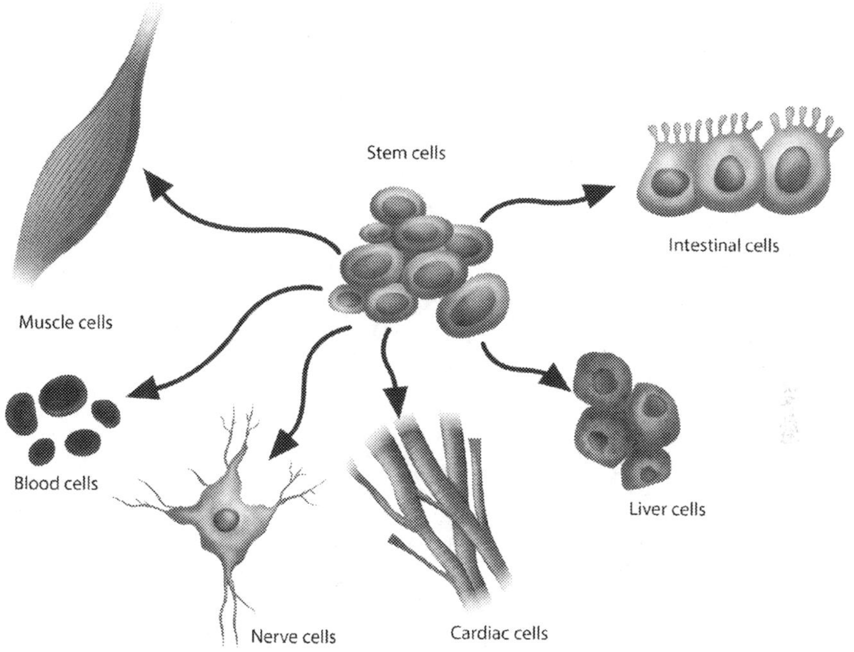

Stem cells can differentiate into multiple types of cells, such as blood cells, nerve cells, muscle cells, cardiac cells, liver cells, and intestinal cells.

typically produce the type of tissue in which they are found. For example, doctors are able to isolate mesenchymal stem cells from adult fat tissue. These are stem cells that develop into fat cells, cartilage, bone, tendons, ligaments, muscle, skin, and even nerve cells. If a mesenchymal stem cell takes another step along the path of differentiation, it can become, for example, a cardiac myocyte, which, as its name implies, produces muscles cells of the heart. Another type of adult stem cell, hematopoietic stem cells give rise to the cell types of the blood system—red cells, white cells, and platelets.

It is these blood stem cells that first provided proof-of-principle for the use of stem cells against human disease. For more than forty years, doctors have been transferring blood stem cells from donors to patients in the procedure known as bone marrow transplant. This use of hematopoietic stem cells is most commonly used to replace bone marrow that has been destroyed by chemotherapy, such as in the treatment of leukemia and lymphoma, and now adult stem cells are being used to fix debilitating autoimmune conditions like rheumatoid arthritis, multiple sclerosis, lupus, and type 1 diabetes, in which a patient's immune system erroneously attacks healthy tissue. In addition to stem cells from an adult donor, cord blood, which is harmlessly collected from umbilical cords that are normally discarded after a baby's birth, is now also used as a source of hematopoietic stem cells for transplant. This use of hematopoietic stem cells has helped to boost cure rates for childhood leukemia from nearly zero in the middle of the twentieth century to greater than 90 percent today.

Now that there is proof that adult stem cells can be isolated and used to cure real human diseases, the race is on to isolate new types of stem cells and match them with the conditions they treat. Call it what you want—regenerative medicine, stem cell treatment, or cell therapy—the result is the same: new hope for healing hundreds of formerly incurable diseases.

Let's look at just a few of the diseases and conditions that are now or may soon be treated with stem cells.

STEM CELLS FOR CHRONIC MUSCLE AND JOINT INJURIES

In 2006 legendary quarterback Peyton Manning injured his neck in a game between his Indianapolis Colts and the Washington Redskins. The star player rehabbed his injury but reinjured it again in 2010. By the 2011 season, it was obvious that pain, and perhaps nerve damage, had sapped his arm strength. Manning then underwent surgery for a herniated disc. When that didn't work, he had vertebrae in his spine fused together. Finally, having exhausted mainstream options, Manning explored other solutions. Details are sparse, because Manning would be a free agent later that year and the last thing he and his agency team wanted was loose talk about experimental and possibly prohibited treatments, but it is reported that during the 2011 season Manning flew to Europe where doctors harvested stem cells from his fat tissue and then injected the purified cells into his neck. What we know for sure is that in 2012, Manning, now with the Denver Broncos, was named the NFL Comeback Player of the Year. In 2013 he was selected the NFL Offensive Player of the Year. In September of that year, in the season opener against the Baltimore Ravens, he threw an NFL-record seven touchdown passes. Two seasons later, Manning led the Broncos to victory in the 2016 Super Bowl, a fitting finale to an illustrious career that was almost sidelined by injury. Was it the stem cells? The treatment he supposedly received certainly coincided with his miraculous recovery, but this kind of correlation doesn't prove that stem cells *caused* the recovery.

In fact, Manning's story illustrates a major challenge in showing the effectiveness of stem cell therapies in muscle, joint, and orthopedic repair. A quick internet search returns pages and pages of patients recounting their stories of miraculous cures, but we have few verifiable facts. A 2013 review in the journal *Knee Surgery, Sports Traumatology, Arthroscopy* sums up current knowledge: "Despite the growing interest in this biological approach, knowledge on this topic is still preliminary, as shown by the prevalence of preclinical studies and the presence of low-quality clinical studies." In other words, scientists have shown in their labs that stem cell injections *should* help repair muscle and joint injuries. And some small clinical trials have hinted that stem cells *do* repair muscle and joint injuries. But until we have the evidence of a large, randomized controlled trial, the United States Food and Drug Administration and most physicians won't be ready to put their stamp of approval on the treatment. That's why Peyton Manning and many other pro athletes go to Europe.

Take Lakers star Kobe Bryant. By 2011, he had worn out most of the cartilage in his right knee, resulting in severe pain that often kept him from practicing. Like Manning, Bryant flew to Europe for treatment, but unlike Manning, his therapy didn't use stem cells. Instead, Bryant opted for injections of platelet-rich plasma (PRP). The process of creating PRP for patients is relatively simple. One or two ounces of the patient's blood are put into a centrifuge where it is spun at very high speed to separate out the platelets. This becomes PRP, a concentrated source of over 300 growth factors that are believed to play a significant role in the healing process. This PRP is then injected into the injured area of the soft tissue or joint to promote healing, possibly by supporting the injured tissue's own dormant stem cells. When this works, most patients experience relief within several weeks and without any side effects. After his PRP session, Bryant told *ESPN.com*, "I feel a lot stronger and a lot quicker."

Of course, pro athletes aren't alone in their desire to heal ailing joints and muscles. A survey by the organization Safe Kids Worldwide found that every year in the United States more than one million kids are injured severely enough when playing sports to require an emergency room visit. And sports aside, the Centers for Disease Control report that about thirty-one million adults suffer from osteoarthritis, the most common form of the disease. We know that mesenchymal stem cells—the adult stem cells that give rise to muscle, bone, and connective tissue—are less active in patients with osteoarthritis. Indeed, an article in the journal *Arthritis & Rheumatology* suggests that the inability of stem cells to keep up with the normal wear and tear on bones and cartilage may be a defining feature of the disease.

Researchers at the Hospital for Special Surgery in New York City have launched a study to determine if a treatment using stem cells can help people with painful osteoarthritis. The multicenter study will divide patients into three groups, with each receiving a one-time injection into the knee joint. One group receives a stem cell treatment. The second group receives an injection of hyaluronic acid, a common treatment that has been shown to help ease pain by lubricating an arthritic joint. The third group receives a placebo that is made up of saline solution and injected into the knee joint. This is

ARTHRITIS OF THE HUMAN KNEE JOINT

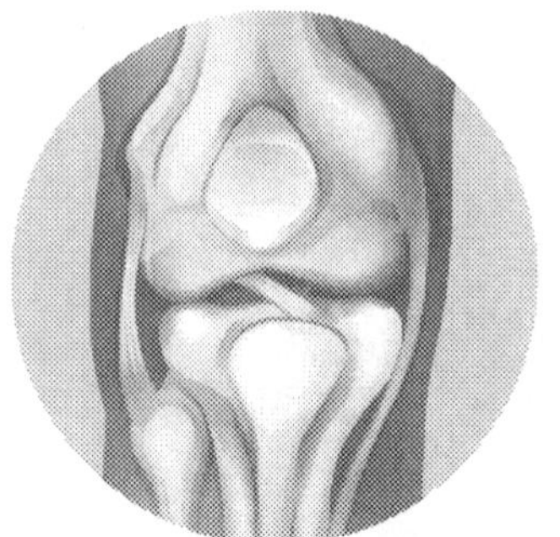

Normal knee joint

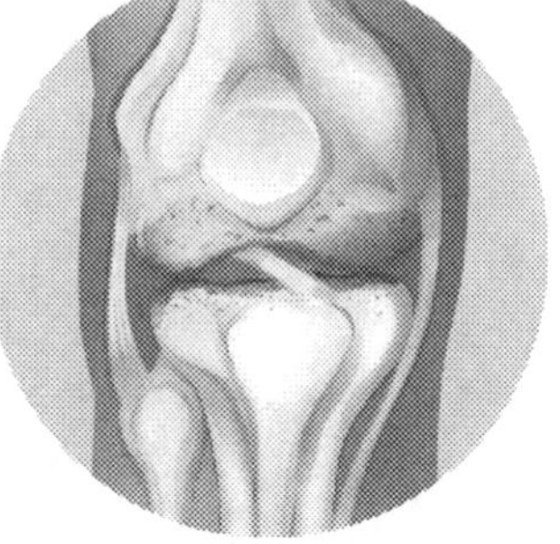

Knee joint with arthritis

Osteoarthritis, the most common type of arthritis, is the result of injury or, most often, wear and tear on the cartilage on the ends of bones so that they no longer glide smoothly. Doctors are working to use stem cells to regenerate the damaged cartilage.

a blinded study, meaning that the participants do not know what is being injected into their sore knees.

"Studies have demonstrated that stem cells are safe and can improve healing and reduce symptoms in a number of different applications, such as cardiac surgery and wound healing," says Sabrina Strickland, MD, an orthopedic surgeon at the Hospital for Special Surgery and principal investigator in the arthritis study.

Study participants will have X-rays taken before the injection and after twelve months. They will also have lab work and will complete questionnaires regarding pain and activity level. Follow-up visits after the initial injection will take place at six weeks, three months, six months, and twelve months.

"I have a number of younger patients with arthritis who are looking for a new pain-relieving option, and the stem cell treatment works in a completely different way from current treatments," Strickland says. "If it is shown to be safe and effective and slows down the progression of mild to moderate arthritis, we'll be able to help a lot of patients."

Further downtown, at New York University's Langone Medical Center, doctors are testing a similar treatment for hip pain associated

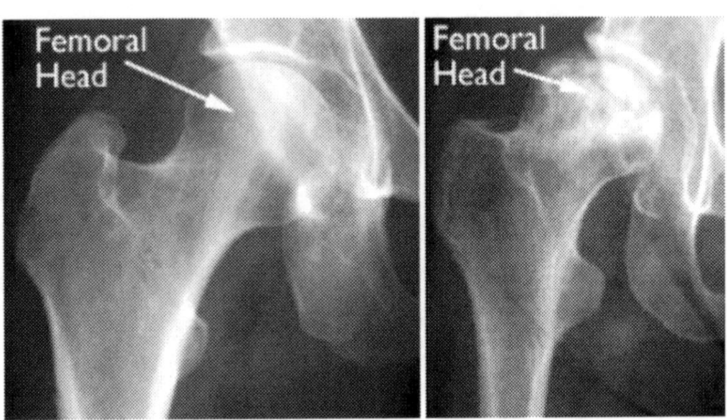

A healthy hip joint (left) has a smooth ball sitting in its socket. When blood supply to the head of the thigh bone is compromised (right), the bone begins to die, becoming mottled and irregular and eventually leading to a total hip replacement unless treatment such as stem cell infusion is done to the bone. (Reproduced with permission from OrthoInfo. ©American Academy of Orthopaedic Surgeons. http://orthoinfo.aaos.org)

with the condition known as avascular necrosis (AVN). AVN, also called osteonecrosis, is a condition in which circulation to the femoral head of the hip—the highest part of the thigh bone—becomes impaired, causing the ball to lose its round shape and eventually collapse and die. The poor fit of the ball into the hip socket damages the joint's surface cartilage, leading to chronic arthritis and eventually destruction of the joint, requiring total hip replacement usually before the age of 50. Thomas Einhorn, MD, professor of orthopedic surgery at NYU and director of clinical and translational research at Langone, has been using adult stem cell therapy to successfully treat AVN in many of his patients, helping them to avoid hip replacement surgery.

"A total hip replacement is often a suboptimal solution for patients under sixty years of age," says Einhorn. "That's because of possible activity restriction and the fact that a synthetic hip will wear out with time. Stem cell therapy could prevent some patients from requiring this operation."

After removing bone marrow from the femur through a thin needle, Einhorn transfers the sample to a device that collects stem cells into a concentrate called bone marrow aspirate concentrate (BMAC). In a one-hour surgical procedure, he then drills out the damaged bone in the patient's hip and injects the BMAC directly into the hip where the damaged bone was removed. The BMAC forms a thick clot that serves as a short-term scaffold—that is, a structure on which new tissue can grow.

"The idea is that once the stem cells are in the cavity, they will induce the development of new blood vessels and start to grow, differentiating into bone tissue," Einhorn says. "This new bone tissue should then use the surrounding necrosed tissue as a scaffold, putting living bone back where it belongs."

Over the years, Dr. Einhorn has had a 65 percent success rate with this innovative cellular procedure, not only decreasing pain and keeping damaged hips from collapsing but allowing patients to avoid hip-replacement surgery.

These are a few examples of the many types of muscle and joint injuries, diseases, and degeneration for which regenerative medicine based on stem cells or on other cell therapies is showing significant results in patients with the means and motivation to undergo experimental treatment. Trials that are currently underway could make these therapies available to thousands more.

MY BROKEN HEART: HEART REPAIR WITH STEM CELLS

Between 1940 and 1967, death by coronary heart disease rose 14 percent in the United States, due in part to smoking, lack of exercise, and high consumption of fatty and processed foods. Since then, the death rate has fallen by almost as much, due to less smoking, better medical techniques, and less time between the heart attack and treatment. But surviving doesn't necessarily mean thriving. Once heart tissue is killed, it cannot regrow and instead is healed over by scar tissue in a process called fibrosis. This process causes the heart to become inflexible and unable to contract and pump blood efficiently. Eventually, an inefficient heart is prone to fail. About 22 percent of men and 46 percent of women will develop heart failure within six years of having a heart attack. Fewer than half of those patients will survive for five years.

But despite the grim statistics, there is hope for better outcomes. Many experts now believe that with stem cell therapy the heart may, in fact, be able to heal dead tissue. This is a huge claim, contradicting what has been an established paradigm in cardiac medicine. Here is the evidence:

In 2009 Eduardo Marban, MD, PhD, director of the Cedars-Sinai Heart Institute in Los Angeles, harvested a small piece of a patient's heart and used the tissue to isolate and grow specialized heart stem cells. Then Marban's team injected these stem cells back into areas of the patient's heart that had been injured by heart attack. Twenty-three other patients joined this first trial, and in 2012 *The Lancet* published the results: one year after receiving the investigational stem cell treatment, heart attack patients demonstrated a significant reduction in the size of the scar left on the heart muscle after their heart attacks. The paper described this regrowth as "unprecedented increases . . . consistent with therapeutic regeneration."

Then, in the largest clinical trial ever conducted using stem cells to treat heart failure, researchers led by Amit N. Patel, MD, a cardiac surgeon and director of the clinical regenerative medicine program at the University of Utah, compared 58 patients who received injections of heart stem cells to 66 patients who received injections of a saline solution (placebo). The treatment, called ixmyelocel-T (developed by Vericel Corporation of Cambridge, Massachusetts), used about three tablespoons of bone marrow from the hip that was then grown in an instrument called a bioreactor to produce a "soup" of cells containing mesenchymal stem cells. Researchers used special catheters to identify the weakest parts of each patient's heart and injected the cellular soup into these areas. The results, reported in *The Lancet*, showed that in the year following the treatment, the 58 patients who received the cell therapy had a 37 percent lower rate of death and hospitalization for heart failure–related problems. Mortality was 3 percent in the patients treated with cell therapy compared to 14 percent in the placebo group.

Patel thinks that the cell therapy probably works not by increasing the number of muscle cells or blood vessels in the heart but by making surviving muscle cells work better. Patel and collaborators at Vericel are hoping to start a much larger phase III trial in the near future. In this trial, the treatment will be given to a large group of people to confirm its effectiveness, monitor side effects, compare it to commonly used treatments, and collect information that will allow the drug or treatment to be used safely.

Patel and Marban are not alone. The Australian company Mesoblast completed a phase II study of mesenchymal stem cells injected into the hearts of patients who had experienced heart failure. "Our results showed that the therapy was not only safe but effective. Six months after treatment, patients had improved heart pump function," says Donna Skerrett, MD, chief medical officer for Mesoblast. At this

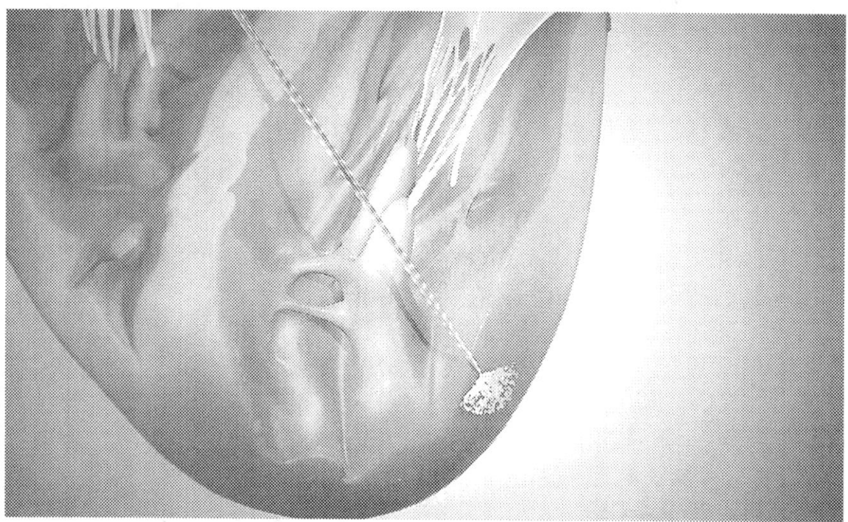

Still from a simulation of Mesoblast's cell therapy, MPC-150-IM, being delivered via catheter to damaged heart muscle. (Courtesy of Mesoblast Ltd.)

writing, a larger, phase III clinical trial with 600 patients is now underway and interim data is expected. Other companies using stem cells to treat heart disease include Celyad, a Belgium-based cell therapy company, which in late 2016 presented promising results of a 271-person study using the patients' own bone marrow stem cells mixed with special growth factors that coax the cells to become cardiac stem cells. This proprietary medicine, C-Cure, has been shown to be effective in a subset of patients selected. And in 2016 the San Diego biotech company CardioCell published promising data from a 22-patient study of their bone marrow–derived stem cell product, which, instead of being injected into the heart, was simply infused into patients' veins. These three studies are examples of more than forty clinical trials by companies such as Fate Therapeutics, Amorcyte Therapeutics, and many others currently testing stem cell–based treatments for cardiac conditions.

The pace of medical research can be frustrating. It can take a decade or more to develop and test a new treatment. However, this snail's pace helps to ensure two essential aspects of research. First, treatments that survive this process are more likely to be safe; and second, this rigorous testing process is the best way to conclusively prove a treatment's effectiveness.

The wheels of scientific research turn slowly. Granted, it's not easy to wait on the sidelines for the results of these trials of stem cells for heart disease by Vericel, Cedars-Sinai, Mesoblast, Celyad, CardioCell, and others. However, if successful, the results of these ongoing investigations will eventually pave the way for stem cell therapy as an accepted treatment for a variety of cardiac conditions.

BUYER BEWARE

Stem cells have enormous potential to help us understand and treat a range of diseases, injuries, and other health-related conditions. However, there are now more than 570 clinics in the United States offering costly stem cell treatments for medical purposes, ranging from spinal cord injuries and Parkinson's disease to dementia and multiple sclerosis. "Unfortunately, many make claims that are not supported by a current understanding of science," says Larry Goldstein, PhD, director of the Sanford Stem Cell Clinical Center at UC San Diego Health.

If you are interested in stem cell therapy for yourself or a loved one, the best place to start is with a clinical trial. Look for trials that are conducted by academic medical centers or nonprofit medical institutions. With most of these experimental studies, you will not be charged for the cost of treatment. Find these studies by using the search tools at the government-run ClinicalTrials.gov database.

CHAPTER TWO

GROWING NEW BODY PARTS IN THE LAB

Regenerative medicine may one day be a solution to the shortage of donor organs in this country for those needing transplants.

—ANTHONY ATALA, MD

Luke Massella is a member of a very small but promising community—people whose damaged or diseased organs have been replaced with new ones grown from their own cells in a laboratory.

Massella was born with spina bifida, a birth defect affecting about 1,500 newborns annually in the United States in which the backbone fails to close completely around the spinal cord. Fortunately, Massella didn't suffer most of the neurological problems that can accompany spina bifida, but the condition did prevent the formation of nerves that help control the bladder. There are two sets of these nerves. One set exits the spinal cord in the middle of the back and controls tightening of the bladder's sphincter muscles so that it can hold urine. The other set, which exits much lower in the back, relaxs those muscles to release urine. Without the proper formation of these nerves, as is the case in the majority of children with spina bifida, the bladder may have trouble storing urine, emptying fully, or both. Over time, bladders that sit dormant become stiff and fibrotic or waste away and eventually die. It was the lower set of nerves that didn't form in Luke.

By the time he was ten, corrosive urine had backed up from his faulty bladder into his kidneys, which were on the brink of failure. Luke had been an energetic boy, but now on most mornings he didn't have enough energy to get out of bed. His bladder was dying. Eventually Luke would die as well. That's when, in 2001, his parents took him to see Anthony Atala, MD, a pediatric urologist who was gaining a national reputation as an expert in the newly developing medical specialty of tissue engineering or, as it is now called, regenerative medicine.

MEDICAL MILESTONE: ENGINEERING A BLADDER

Dr. Anthony Atala proposed a solution that seemed like science fiction: use Luke's own stem cells to grow a completely new, functional bladder. Here's how Atala explained the procedure in a TED talk about his work: "We took a very small piece of tissue from his injured organ, a little less than half the size of a postage stamp. We then took those cells and grew them outside his body in my laboratory. A 3D biodegradable mold in the shape of a bladder was created, and then we soaked and seeded it with nutrients and the stem cells from Luke's own bladder."

Atala explained that tissue was built up in layers and then incubated in a special bioreactor, rather like baking a layer cake. "When we were done, the new bladder was then surgically implanted into Luke and attached to his original bladder."

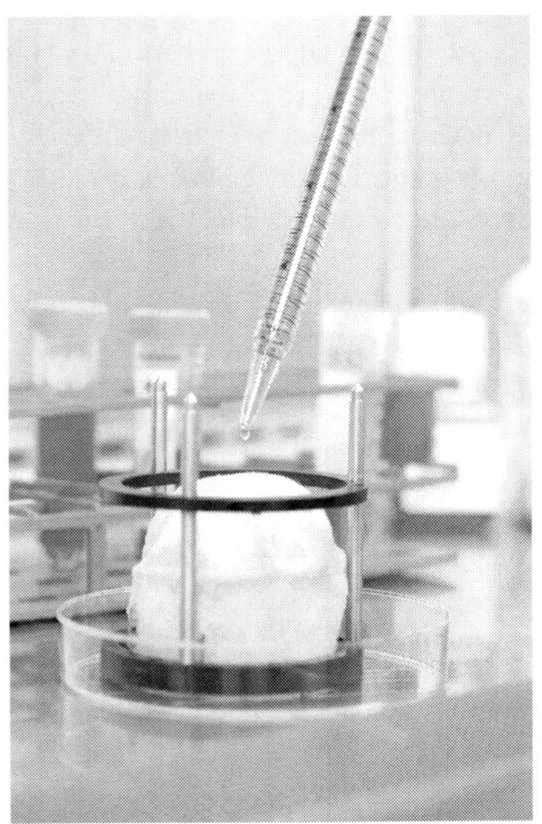

A bioprinted bladder scaffold about to be seeded with the patient's stem cells grown in the lab of Dr. Anthony Atala. (Courtesy of Wake Forest Institute for Regenerative Medicine)

Soon this custom-built organ attracted its own blood supply and even encouraged the growth of new nerves, neatly wiring itself into the ecosystem of Luke's body. The impact on Luke's life was greater than you can likely imagine. Seven years after his surgery, Luke was selected as the captain of his high school wrestling team, and after graduating from the University of Connecticut, he went on to become a high school wrestling coach. "This bladder has allowed Luke to have a much better life," says Atala, in an obvious understatement. The new bladder saved Luke's kidneys, and now, more than fifteen years after the procedure, his made-to-order bladder continues to function.

FROM SALAMANDERS TO HUMANS

In the early 1990s, while working as a postdoc at Harvard, Atala read an article about salamanders that asked a simple question: Why is it that salamanders can regrow lost limbs but humans cannot? The question became a fascination, leading Atala to explore the inner workings of the human system of healing. The problem, as he saw it, was that healing was simply too slow. We have the capability to heal cuts and regrow bone, but the process is too slow to offset major injuries and diseases—so much that the result is often a net loss rather than a net gain. Now, at the Wake Forest Institute for Regenerative Medicine in Winston-Salem, North Carolina, Atala leads an interdisciplinary team of 450 biologists, chemists, engineers, and support personnel united by a common goal: to leverage his understanding of the healing powers of the body's own cells to regenerate damaged tissues and organs.

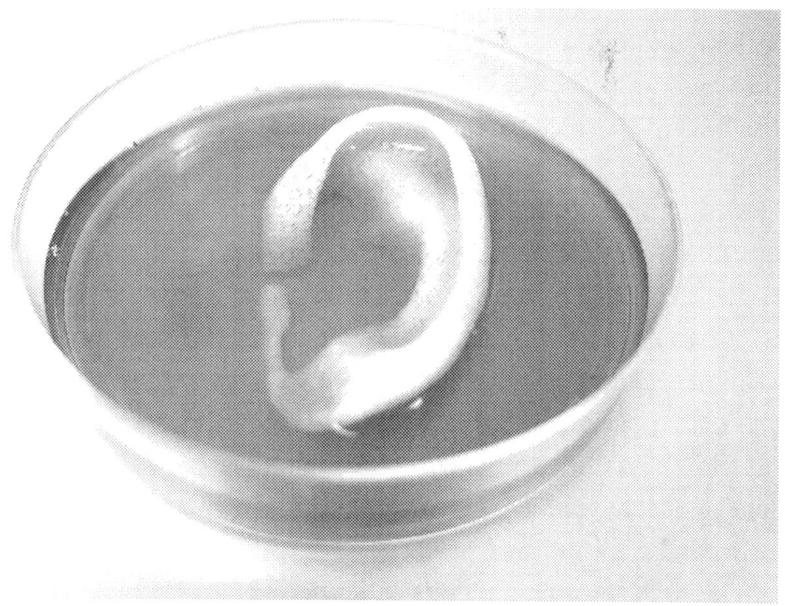

A bioprinted ear scaffold being seeded with stem cells in preparation for grafting onto the back of a lab mouse to demonstrate viability.

"At the institute, we are now growing over thirty different types of tissues and organs, some that we already have put into patients," Atala says. Arteries, ears, finger bones, urethras, kidneys, tracheas, artificial heart valves, tiny muscles, and more are being grown in the lab at Wake Forest.

One of Atala's engineered tissues is making a difference for girls born with Mayer-Rokitansky-Küster-Hauser Syndrome (MRKHS). This emotionally and physically painful disorder, which affects about one in 5,000 women, occurs when the uterus and vagina fail to develop properly, resulting in a vaginal canal that is short and narrow or even absent. The rest of the external genitalia, as well as ovarian function, are not affected. Women generally realize they have the condition in their mid to late teens, when their periods don't arrive or when attempts to have intercourse are painful and difficult.

Between 2005 and 2008, in an effort to help women with this condition, Atala implanted new lab-grown vaginal organs in four girls with MRKHS, ages thirteen to eighteen. The organs were grown with a technique similar to the one Atala used for Luke Massella's replacement bladder. First, Atala's team extracted muscle and epithelial cells from a small sample of each patient's external genitals. These cells were grown in the lab and then sewn onto a biodegradable scaffold that was handcrafted to fit each patient's anatomy.

About six weeks later, surgeons created a canal in each patient's pelvis and sutured the scaffolds to the girls' reproductive structures. As in Luke Massella's bladder treatment, once the cell-seeded scaffolds were implanted, nerves and blood vessels grew in around them, and the cells expanded and organized themselves into tissue. As the cells grew to form their permanent structure, the girls' bodies absorbed the biodegradable scaffold so that gradually the engineered scaffold was completely replaced by a new organ.

In 2014 Atala reported his patients' long-term success in *The Lancet*, showing that even up to a decade after their surgeries, the organs continued to function. In addition to the tissue remaining viable, his patients' responses to a Female Sexual Function Index questionnaire showed that they had normal sexual function after the treatment, including arousal, pain-free intercourse, and orgasm.

"This may represent a new option for patients who require vaginal reconstructive surgeries," says Atala. "The treatment could also potentially be applied to patients with vaginal or cervical cancer, or injuries suffered in a car accident." It could also become an option for men seeking gender reassignment. Importantly, Atala's lab-grown organs are made from patients' own cells, dramatically reducing the chance that a patient's body will reject the new organ.

PRINTING BODY PARTS

A major challenge in reaching more patients with Atala's techniques is the fact that these procedures have to be reproducible—that is, standardized procedures are needed that can be followed by doctors in other labs and clinics. The problem was that the work being done in Atala's lab was as much art as science; the people constructing these tissues and organs were essentially sculptors. This challenge led Atala to explore more step-by-step ways to shape scaffolds and seed these scaffolds with cells.

The result is his Integrated Tissue and Organ Printing system (ITOP), a machine that literally prints tissues and organs by using a combination of biodegradable plastic sheets and water-based gels that contain a patient's cells. Basically, doctors use CT or MRI scan images as blueprints to determine the shape and dimensions of new organs for each patient. Then they harvest and grow cells. In the actual printing process, the ITOP system lays down a sheet of paper-like biofilm and prints upon it layers of cells in the shape determined by the scans. Layer upon layer, these sheets form the 3D structure of the new organ. The cells then grow through these sheets until eventually the biofilm is gone and an organ composed of a patient's living cells remains.

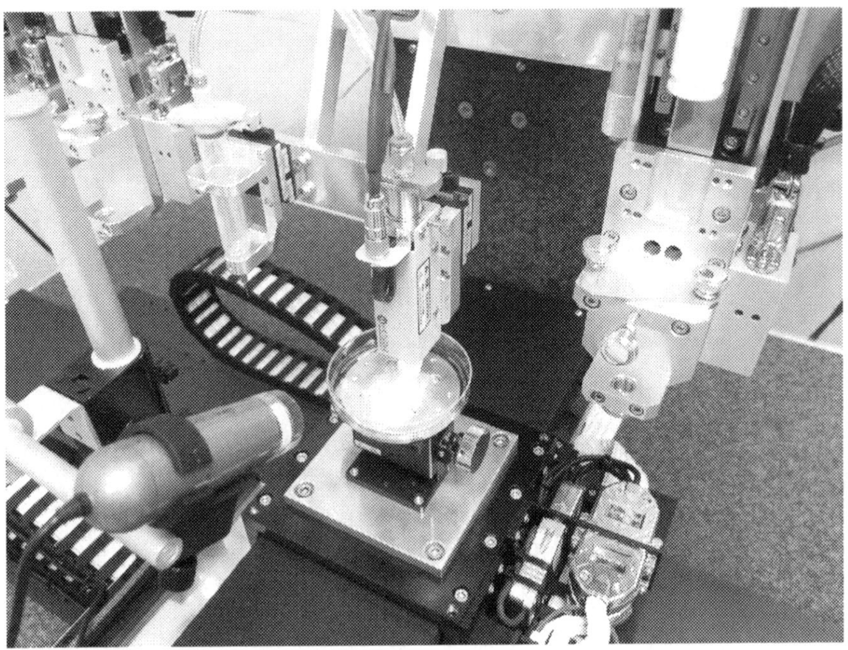

Dr. Atala's 3D-printing system prints tissues and organoids.

Unfortunately, 3D-printed organs are not yet in human use. "When we do our research, our motto is 'first, do no harm,'" Atala says. "When we finally have the science, the data, and the trust, we will be able to go to our first in-man clinical trials with printed tissues and hopefully then we will deliver on our promise. Our goal is to have them available to patients as soon as possible."

That said, a 2016 article in the journal *Nature Biotechnology* describes several of Atala's complex 3D structures that have been successfully implanted in rats and mice. In one experiment, Atala's team embedded printed, human-sized ears under the skin of mice. This made for singularly unattractive mice, but the experiment worked. After two months, the shape of the implanted ear was well maintained, and cartilage, tissue, and blood vessels had formed, integrating the ear into the mouse's body. In another experiment, printed muscle tissue was implanted in rats. Two weeks later, tests confirmed that the muscle was robust enough to maintain its structure, and again the tissue had been surrounded by new blood vessels and nerves. The same was true with bone. Atala's team used human stem cells to bioprint fragments of bone like those needed for jaw-reconstruction surgery. Five months after these fragments were implanted in rats, they had formed vascularized bone tissue.

The longer we live, the more likely we are to experience organ failure. Ergo, as the average human life span increases, so does the need for organ replacements. Increasing the supply of donor organs is problematic; even in the case of kidneys, where we don't have to wait for someone to die in order to harvest an organ, there aren't nearly enough donor organs available, and people are dying in need of them. Atala's work offers the promise that we may be able to fill this shortage safely and reliably by growing and printing replacement organs in laboratories.

REGENERATING THE ESOPHAGUS

Dr. Anthony Atala isn't the only medical pioneer advancing our ability to generate human organs, nor are bladders and reproductive structures the only replacement tissues that have been used in humans. Each year, approximately 17,000 Americans are diagnosed with esophageal cancer, and about 15,000 of them will die from the disease. Accidents and infections destroy thousands more of these essential tubes, which carry foods, liquids, and saliva from mouth to stomach. In extreme cases, the esophagus must be extracted, a surgical procedure in which part or all of the esophagus is removed and replaced by molding the remaining stomach organ into a tube. Now, thanks to cellular technology, there may be a better, less traumatic way to repair this damage.

An astonishing example is the case of a twenty-four-year-old man that was reported in a 2016 article in *The Lancet.* In 2009, he had been admitted to a hospital emergency department with a life-threatening esophageal infection. By the time the young man sought care, his esophagus was already too damaged to be repaired through a surgical reattachment procedure. With no good options remaining, the hospital reached out to Kulwinder S. Dua, MD, a gastroenterologist from the Medical College of Wisconsin in Milwaukee, who had shown success in dogs with a dramatic, experimental procedure.

The more desperate a patient's case, the more likely the FDA is to approve an unproven, radical treatment, and that's exactly what happened. With the young man's life hanging by a thread, Dua and his team were granted permission to operate. They used metal stents to build a scaffold bridging an essential, two-inch section of the man's esophagus. Dua then populated the scaffold with cells from the man's own muscle along with donated skin, and then "fertilized" the tissue with platelet-rich plasma spun from the patient's blood. Here's an important point: in addition to *making* new tissue, the techniques of regenerative medicine can also stimulate the body's own ability to repair and regrow tissue. That's what happened here. Not only did Dua reconstruct tissue through surgery, but the patient's cells and platelet-rich plasma encouraged his body to saturate the new tissue with growth factors and also attracted stem cells that helped the esophageal bridge to heal.

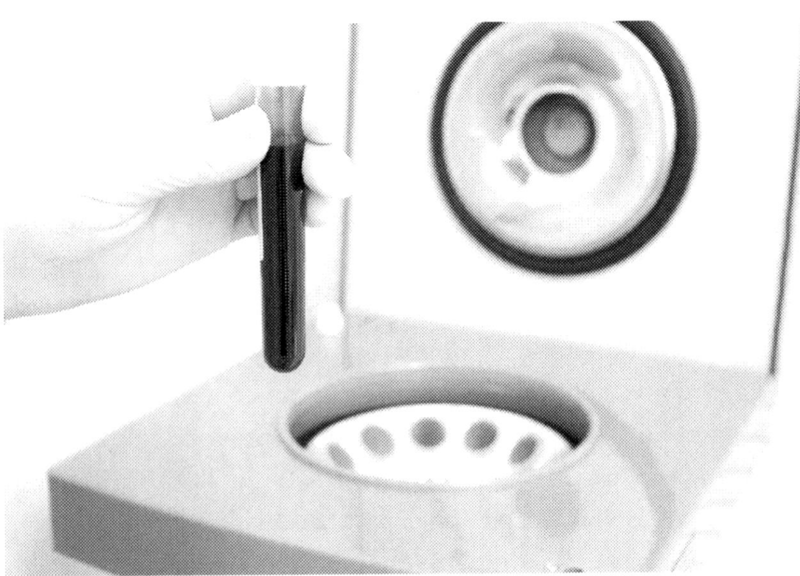

Centrifuge for separating platelet-rich plasma from blood.

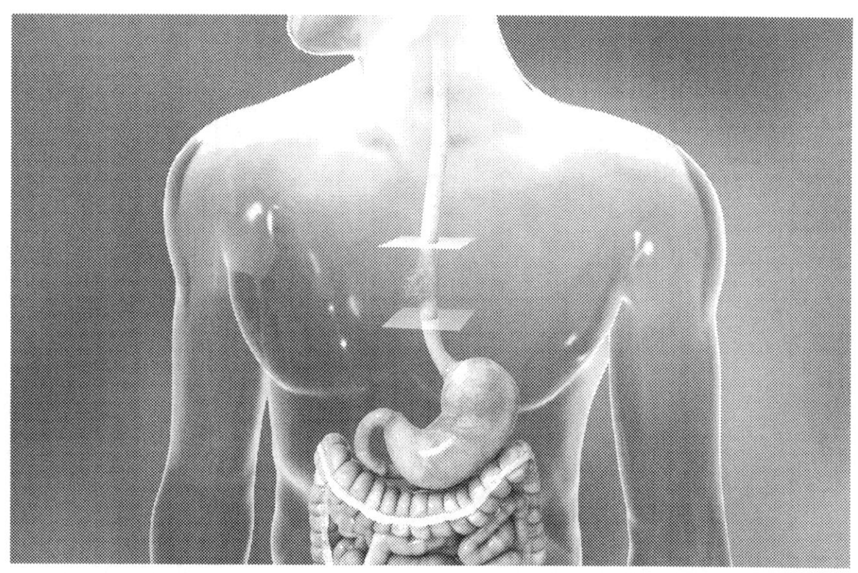

Diagram showing an area of diseased esophagus (box), often a tumor or infection, requiring that section of food tube to be surgically removed. But now there is not enough length of esophagus to connect to the stomach. (Courtesy of Biostage Inc.)

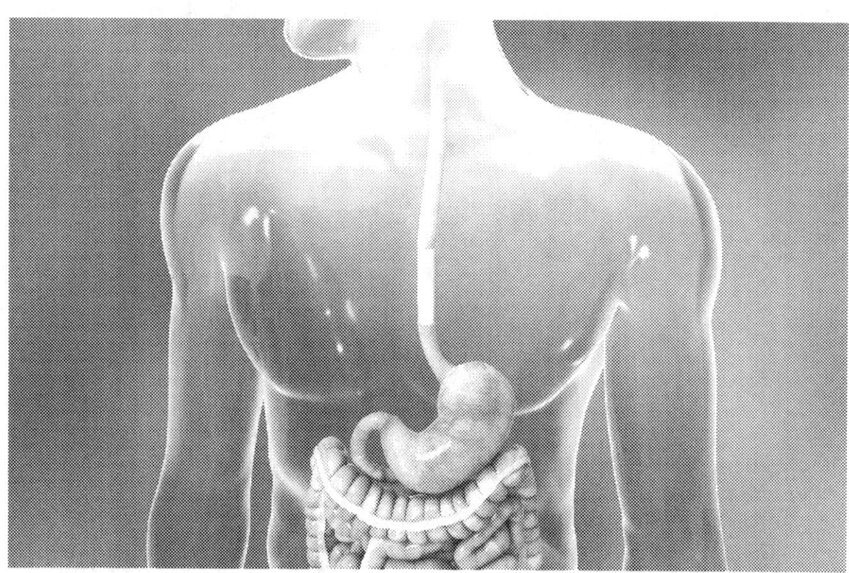

Doctors have been able to grow a length of replacement esophagus by seeding a biodegradable scaffold with stem cells and using it to bridge the gap where the diseased part of food tube was removed. (Courtesy of Biostage Inc.)

Four years following the surgery, Dua and his team removed the patient's stents. A year later, follow-up tests revealed regeneration of all five layers of the esophageal wall, while tests showed that, amazingly, the patient could swallow normally. His new esophageal muscles had learned to constrict in the complex waves known as peristalsis that massage food down the pipe. To date, the patient continues to eat a normal diet, has no problems swallowing, and maintains his weight.

Based on this initial success, Dua believes that regenerative medicine and stem cells could profoundly change the way esophageal damage and disease are treated in hospitals around the world. To make the transition from a one-time use in an extreme case toward more widespread use, Dua has applied for grants to launch animal studies, followed by the phase I and II clinical trials needed to get FDA approval for the cellular treatment. Meanwhile, researchers at the Mayo Clinic, in collaboration with the Holliston, Massachusetts, company Biostage, have refined the technique by using a biodegradable scaffold so that patients do not require another surgery to remove the lattice-like framework

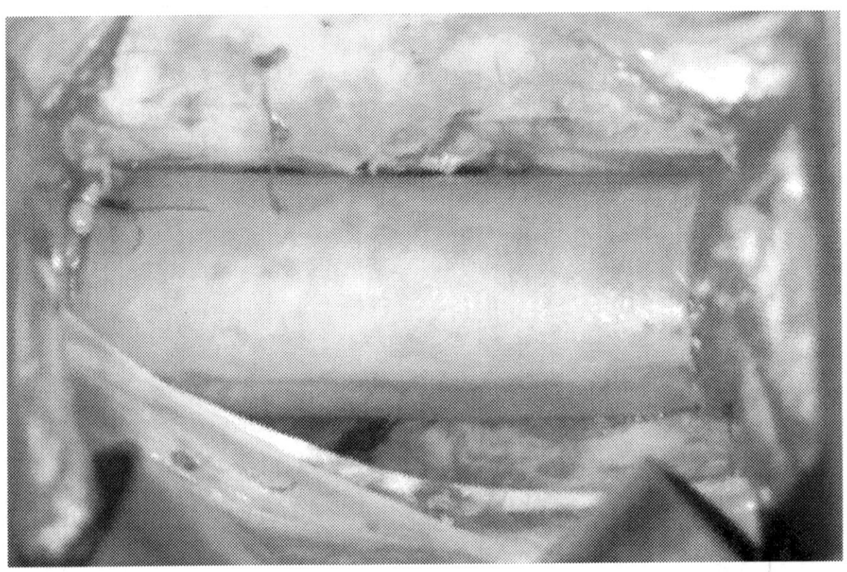

An esophageal scaffold, seeded with stem cells, sewn in place in a patient's chest. (Courtesy of Biostage Inc.)

that supports the growth of stem cells into tissue. Biostage applied to the FDA to start human trials by 2017 and hopes to show that a bioprinted esophagus can function normally in eating and swallowing.

STEM CELLS FOR EXPERIMENTAL MODELS

We're all amazed by the idea of a new organ grown with stem cells and implanted into a patient's body. But many scientists think that an equally important application of regenerative medicine will be to create accurate, humanlike models that researchers can use to test new treatments. For obvious reasons, medical researchers can't poke, prod, and medicate a human heart to see what happens. Instead, scientists experiment with cells in a dish or with nonhuman animal models. Unfortunately, what works in a dish does not always work with humans, and just because drug safety is established in mice doesn't mean a new compound won't cause problems in human trials. Every year, academic research institutions and pharmaceutical companies throw away billions of dollars and huge amounts of time on drugs that seem promising in cell and animal studies but fail in late-stage human trials. The resources invested in drugs that fail could have been used to speed the approval of effective, lifesaving medicines.

For example, in 2015 researchers from Murdoch Childrens Research Institute in Brisbane, Australia, reported in *Nature Cell Biology* that they had grown a minikidney the size of that in a thirteen-week fetus from induced pluripotent stem cells (iPSCs) taken from adult skin. Though growing a functional kidney large enough for transplant in a human is likely more than a decade away, the San Diego–based company Organovo has already licensed the Australian technology to grow minikidneys for use in drug testing. Lab-grown kidneys may enable researchers to more accurately test the safety and efficacy of an experimental drug before investing in costly human trials. Organovo is also using its own 3D-printing technology to make liver tissues for use in drug testing. In fact, seven of the top twenty-five major drug companies are now testing their drugs with Organovo's bioprinted tissues.

Researchers from the Max Planck Institute for Infection Biology in Berlin are using stem cells to make models of fallopian tubes that mimic those found in human beings. Fallopian tubes are the structures in a woman's body that conduct eggs from the ovaries to the uterus. Unfortunately, they are prone to blockage from infection and are also implicated in the development of some cancers. Reporting in the journal *Nature*, the German scientists describe how they removed epithelial cells with potential stem cell properties from fallopian tube samples and placed them in special growth-promoting conditions. As the cells grew, the scientists noticed that some were coalescing into organoids—small structures made of thousands of cells that have the same shape and structure as the original fallopian tube. The team's miniature fallopian tubes have now survived in the lab for more than a year, during which time the scientists have used these models to discover two important signaling pathways that activate stem cells for tissue repair.

Regenerative medicine has developed at a lightning pace, so much so that scientists are now experimenting with the creation of more complex organs—not just tubes, bags, and homogeneous tissues, but organs with structure and function. At the top of the wish list is a working model of the human heart. Researchers at the University of California, Berkeley, and Gladstone Institutes in San Francisco aren't far off, having developed micro heart muscles in their labs. Reporting in the journal *Scientific Reports* in 2016, the researchers describe how they use these cells, derived from iPSCs made from patients' skin cells, to make 3D human heart tissue. Earlier models using a similar strategy require far more iPSCs and result in an immature heart, more like that of a fetus than an adult's, making it inappropriate for drug testing. To make more mature hearts using far fewer cells, the teams at University of California, Berkeley, and Gladstone Institutes used an innovative new technique; after they generated heart muscle cells and connective tissue cells from iPSCs, they combined them in a special dish shaped like a tiny dog bone. This unique shape encouraged the cells to self-organize into elongated muscle fibers. Within a few days, the micro tissues resembled heart muscle both structurally and functionally, optimized for drug testing or screening to gauge the impact of medications on the heart.

Lungs, too, have been modeled using regenerative cellular techniques. Like the heart, it is not only the composition of a lung cell but the 3D structure of the lung itself that creates its function. Still, for decades researchers have been making do with 2D models—thin layers of lung cells grown in cell-culture dishes. Unfortunately, such models are usually only useful for very preliminary analysis.

That all changed in 2015, however, when researchers at the University of Michigan along with colleagues from the University of

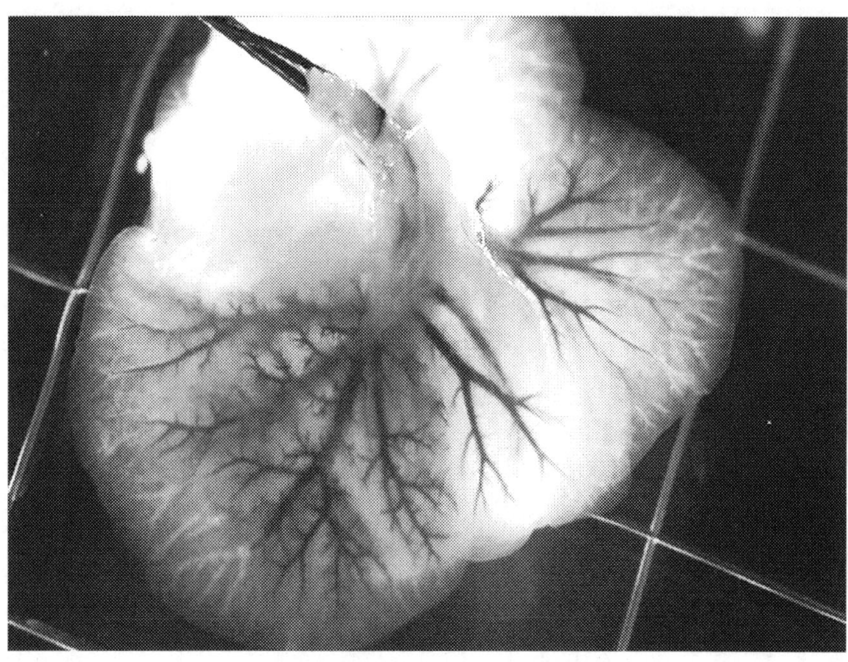

Liver organoid in petri dish. Blue dye fills blood vessels. Organoids like the one in the photograph can be used to test disease treatments. The technology may eventually be used to grow full-scale organs for replacement in humans.

California, San Francisco, Cincinnati Children's Hospital Medical Center, Seattle Children's Hospital, and the University of Washington announced that they had grown the first 3D minilungs from stem cells. First, they "instructed" adult stem cells to form a type of tissue called endoderm, which, as you may remember from this book's first chapter, is one of the three original germ-layer tissues of early embryonic development. The scientists then activated two important developmental pathways that are known to make endoderm form 3D tissue. By inhibiting two other key developmental pathways at the same time, the researchers caused the endoderm to become tissue resembling the early lung tissue found in embryos.

Scientists salivate at the prospect of these stem cell–based models. But in the end, it's patients who benefit. The ability to pretest the safety of new drugs means that patients in clinical trials are more likely to benefit from these exciting new treatments. Weeding out unsafe drugs earlier in the developmental process means that drugs that do make it to the human testing phase will be safer and likely more effective. Likewise, the ability to test hundreds or thousands of drug candidates against humanlike tissue and organoids will result in more efficient development of more effective therapeutic compounds.

In short, the techniques of regenerative medicine benefit patients both by supplying new organs and tissues and by speeding up the process of drug development and lowering the cost. The new medicines and treatments that emerge from both pathways will help people stay healthier longer, and that's the *real* goal of regenerative medicine.

CHAPTER THREE

REPAIRING THE BRAIN

If the brain sows not corn, it plants thistles.
Money spent on brain is never spent in vain.

—ENGLISH PROVERB

On March 4, 2016, thirty-five-year-old Sonia Olea Coontz of Long Beach, California, posted an ultrasound image to her Facebook page. The image showed a beautiful developing baby, Sonia's first child. In one picture, the hemispheres of the baby's brain look like cupped hands or like the lobes of a heart-shaped leaf. The photo followed one posted just over a year earlier in which Sonia announced that she had married her best friend, Peter. For Sonia, this Facebook timeline fits with what you might expect from the natural rhythm of life. You would never know that in 2011 she nearly died from a stroke that left her broken and barely able to speak.

"I could only move my right arm very little," she said in an interview with Stanford University, "and I was in a lot of pain. Same with my leg. Walking was very difficult. Every time I went to the hospital I was in a wheelchair because it was just a lot easier. And speaking was hard. I always needed someone to help me communicate."

In fact, Sonia had gotten engaged just before her stroke, but now she didn't want to be married as she was too embarrassed by her inability to walk down the aisle. One morning six months after her stroke, she woke up with her right arm pinned painfully beneath her. She couldn't seem to pull it free and wanted to call her fiancé for help but realized that she couldn't form the words.

Sonia is one of the 800,000 people in the United States who suffered a stroke or brain attack that year. Stroke occurs when an artery that supplies blood to part of the brain becomes blocked or ruptures. In ischemic strokes, a blood clot is to blame, while a hemorrhagic stroke is typically caused by a burst blood vessel. When blood flow is interrupted to a portion of the brain, neurons become starved of nutrients and oxygen. Damage occurs within minutes, sometimes resulting in irreversible destruction.

Every four minutes, an American dies from stroke complications, making stroke the fourth leading cause of death in this country.

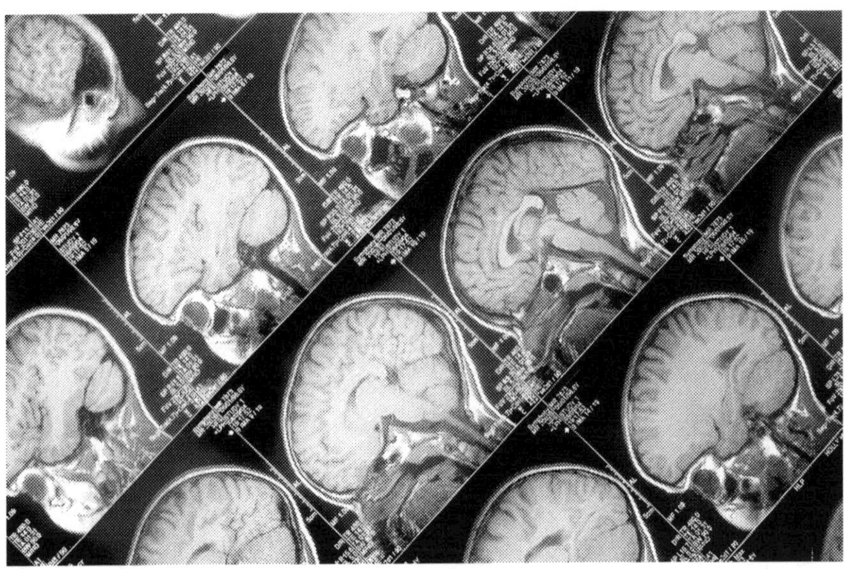

TYPES OF HUMAN BRAIN STROKE

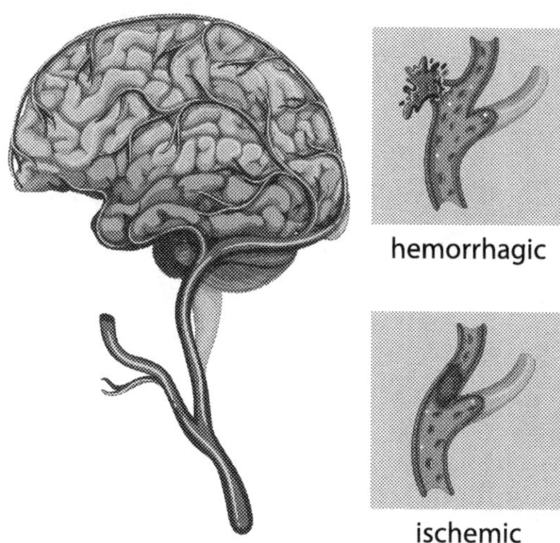

Although many people survive a stroke, even multiple strokes, many suffer permanent neurological damage. Stroke is the number one cause of disability in the United States.

However, the surprising results of a small clinical trial give hope to patients living with the consequences of stroke. In 2015, investigators at the Stanford University School of Medicine and the University of Pittsburgh injected modified human adult stem cells directly into the brains of chronic stroke patients. All 18 patients in this trial (12 at Stanford and 6 at Pittsburgh) had measurable improvements. One study participant, a 71-year-old woman, started the trial able to move only her left thumb. In the weeks following her stem cell procedure, however, she was able to lift her arm above her head. And then she stood up from her wheelchair and walked.

The patients, all of whom had suffered their first and only stroke between six months and three years before beginning the study, remained conscious under light anesthesia throughout the procedure. After drilling a small hole through the skull, doctors injected mesenchymal stem cells—the kind that develop into, among other things,

STROKE

A stroke, or a brain attack, occurs when an artery that supplies blood to part of the brain becomes blocked or ruptures. As a result, blood flow is interrupted to a portion of the brain, and nerve cells (neurons) in the affected area are starved of the oxygen and nutrients they need to function properly. In as little as a few minutes, these nerve cells can suffer damage. After a few hours of interrupted blood flow, the cells can't survive and some brain function is lost.

795,000 SUFFERERS A YEAR IN THE US
600,000 first attacks
185,000 recurring attacks

AGE
Can occur at **ANY** age
3/4 of strokes occur after **65**
After **55** the risk of having a stroke more than **doubles** every decade

DEATH
Fifth leading cause with **130,000**
Death rates are **higher** for African Americans than for whites, even at younger ages

EVERY 40 SECONDS
Someone suffers a stroke in the United States
Someone dies from a stroke **every 4 minutes** in the United States

DISABILITY
LEADING CAUSE
of serious, long-term disability in the United States

TYPES OF STROKE
HEMORRHAGIC AND ISCHEMIC
Hemorrhagic: An artery in the brain leaks blood or ruptures (breaks open), depriving brain cells of blood flow
Ischemic: A blood clot or mass blocks a blood vessel, cutting off blood flow to a part of the brain
87% of strokes are **ISCHEMIC**

RISK FACTORS
HIGH BLOOD PRESSURE: Most important risk factor for stroke
SMOKING: Risk of ischemic stroke in current smokers is double that of nonsmokers after adjustment of risk factors
ATRIAL FIBRILLATION: Independent risk factor for stroke, increasing risk about **fivefold**

US Centers for Disease Control and Prevention

nerve cells—directly into the areas of the brain that were affected by each patient's stroke. The next day they all went home.

What study investigator Gary Steinberg, MD, PhD, the chief of neurosurgery at Stanford University School of Medicine, did not anticipate was the significant—and unprecedented—degree of improvement in all 18 patients within just one month. Some were lifting their arms, getting out of bed, and walking unassisted—things they hadn't been able to do in the months or years since their strokes.

According to the study results published in the journal *Stroke*, more than three-quarters of the volunteers suffered from transient headaches afterward, but those could have been linked to the injection surgery. More importantly, there were no side effects attributable to the stem cells themselves and no life-threatening adverse effects linked to the procedure used to administer them.

Steinberg is not sure how the stem cells worked their magic with these patients. He has postulated that the stem cells may jump-start neuronal circuits that were once thought to be irreversibly damaged or dead. Other experts have suggested that the stem cells may secrete chemicals (or signal nearby cells to secrete chemicals) that activate brain cells so that they begin to function in place of the damaged parts of the brain.

Interestingly, the implanted stem cells themselves do not appear to survive very long in the brain. Preclinical studies have shown that these cells begin to disappear about one month after the procedure and are gone by two months. Yet, while patients showed significant recovery during that first month, their progress continued after the stem cells died. In fact, patients continued improving for several months afterward, sustaining these improvements at six and twelve months after surgery—well after the injected stem cells had faded from their brains.

Sonia was one of the lucky dozen patients who received this treatment at Stanford. "After the surgery, I was immediately better," she said. It was the simple improvements she appreciated most, like regaining the ability to call a restaurant to make dinner reservations. She said that her arm felt about 60 percent improved, and her leg had healed to the point that instead of using a wheelchair she could run and drive. In early 2017 her Facebook header showed her husband, Peter, snuggling their young son.

"Everything is good," Sonia said, "and it's just going to get better."

ALZHEIMER'S DISEASE AND STEM CELLS

Stroke is not the only target in the brain for regenerative medicine. A study published in the journal *Alzheimer's Research & Therapy* found that between 2002 and 2014, 99.6 percent of all experimental therapies directed at Alzheimer's disease had failed, despite the billions of dollars spent on research. To date there is no approved treatment that reverses, stops, or even effectively slows the progress of this disease, which the Alzheimer's Association estimates affects 5.4 million Americans. There are five drugs on the market that treat Alzheimer's symptoms after the patient has already been diagnosed, but they only serve to delay the inevitable. Four of the five drugs are cholinesterase inhibitors, which can slow the symptoms of confusion and memory loss, but these offer only a temporary solution, as patients continue to deteriorate, on average, after six to twelve months. As the number of Alzheimer's patients continues to grow, we are reminded that we are in a race against time. The association estimates that by 2020 more than 40 percent of adults over the age of eighty will have been diagnosed with Alzheimer's.

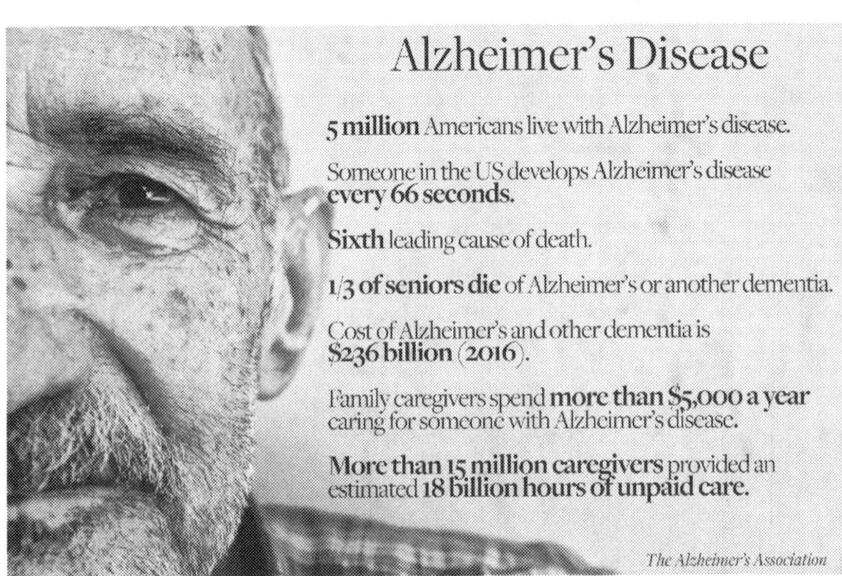

Alzheimer's Disease
5 million Americans live with Alzheimer's disease.
Someone in the US develops Alzheimer's disease every 66 seconds.
Sixth leading cause of death.
1/3 of seniors die of Alzheimer's or another dementia.
Cost of Alzheimer's and other dementia is $236 billion (2016).
Family caregivers spend more than $5,000 a year caring for someone with Alzheimer's disease.
More than 15 million caregivers provided an estimated 18 billion hours of unpaid care.
The Alzheimer's Association

California Institute for Regenerative Medicine (CIRM) created twelve major stem cell research facilities and has given over 800 research grants in the state using $3 billion in funding over ten years. (Courtesy of California Institute for Regenerative Medicine)

However, scientists are increasingly optimistic that stem cell treatments yet to be approved may hold hope for these patients.

"I am a strong and vocal stem cell proponent," says Larry Goldstein, PhD, Distinguished Professor in the Department of Cellular and Molecular Medicine and the Department of Neurosciences at the University of California, San Diego, School of Medicine, and one of the world's foremost experts in stem cell research. Goldstein helped craft Proposition 71, the California Stem Cell Research and Cures Act. This important 2004 law led to the creation of the California Institute for Regenerative Medicine, which distributed $3 billion over ten years to fund stem cell research in the state and continues as a world leader in the field.

Goldstein admits that finding a drug to slow the onset of Alzheimer's has been frustratingly elusive. Humans are the only species that naturally develops Alzheimer's, and confusion about the exact cause of the disease makes it difficult to artificially model the full condition in nonhuman animal subjects. One solution has been Goldstein's use of stem cells to grow petri dish models of the disease.

"These experiments with cells grown in my lab have been extremely helpful in allowing me to understand how various genes work normally and, conversely, what happens when they have a mutation that triggers Alzheimer's," Goldstein says. "In the lab, we can use stem cells to generate with good accuracy any cell type we want to study. For example, to better study the disease, we are using pluripotent stem cell lines—the stem cells that can be engineered to grow into many different tissue types—containing known mutations that cause a type of hereditary Alzheimer's disease. This important work is helping me figure out what's gone wrong with the brain impacted by Alzheimer's disease and why neurons get sick and die."

Another approach has been to use mice to model specific features of the disease—not necessarily Alzheimer's but, for example, the buildup of amyloid plaques in the brain that is associated with the condition. Using human stem cells to generate hereditary Alzheimer's in petri dish models has increased our understanding of the disease; using stem cells with these amyloid-plaque mice may help us learn to treat it. When

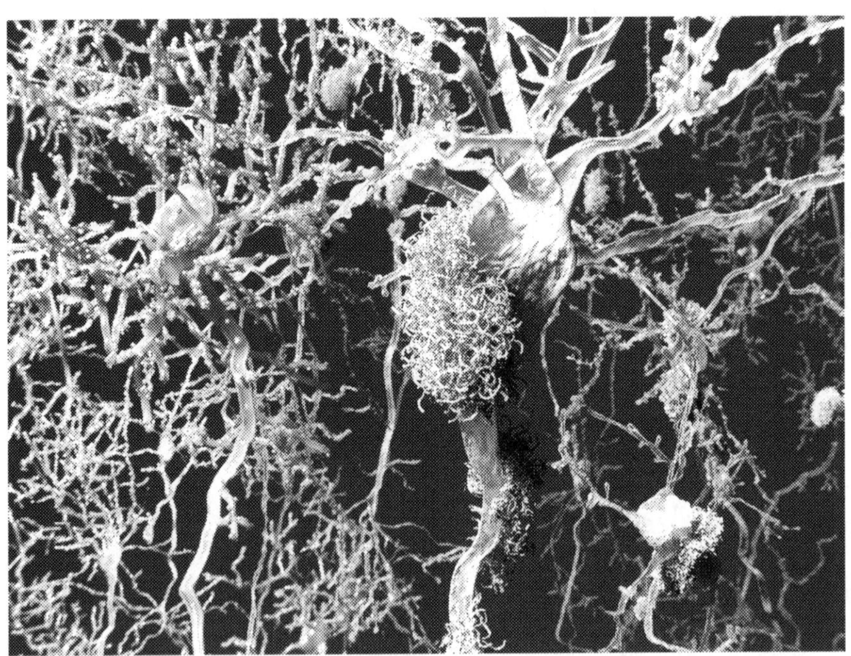

Amyloid plaques are sticky protein buildups that accumulate outside brain cells in Alzheimer's disease. The beta form of amyloid is toxic to neurons, but it's not clear whether it is the cause of Alzheimer's or an "innocent" by-product of the actual disease process.

researchers injected stem cells into the brains of these mouse models, they watched the plaques recede and the mice regain their ability to learn and remember the best paths through mazes.

THE FIRST HUMAN STEM CELL TRIAL FOR ALZHEIMER'S DISEASE

The promise of this work in Goldstein's lab and elsewhere has given FDA regulators enough confidence in the safety and possible benefit of this treatment to approve the first human trial of stem cells against Alzheimer's. The trial has quite a name: the Allogeneic Human Mesenchymal Stem Cell Infusion versus Placebo in Patients with Alzheimer's Disease study, or, we could say, the AHMSCIVPIAD study.

Even without a catchy abbreviation, this first in-human study is getting a lot of attention. Like any phase I study, this one is small, with the goal of exploring treatment safety. Scientists would also "like to see that it slows down the progression of the disease," says principal investigator Bernard S. Baumel, MD, assistant professor of neurology at the University of Miami Miller School of Medicine.

Baumel and his team are using mesenchymal stem cells harvested from donated umbilical cord blood. These cells, extracted from umbilical cords that are saved and banked when a baby is born, are able to grow not only copies of themselves but also a range of tissues, including bone, muscle, cartilage and, importantly, the precursors of nerve cells.

As in the Stanford trial of stem cells in stroke patients, Baumel's hope is not that these mesenchymal cells themselves become new brain neurons in Alzheimer's patients but that they facilitate the healing and growth of the brain's existing cells, or perhaps stimulate the brain's own native dormant stem cells to "wake up" and replenish those that have been damaged by Alzheimer's.

Injections of mesenchymal stem cells have been shown to help control nerve damage as well as support the brain's own healing. They've

ALZHEIMER'S DISEASE

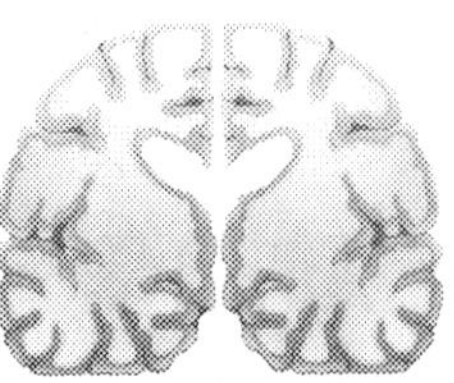

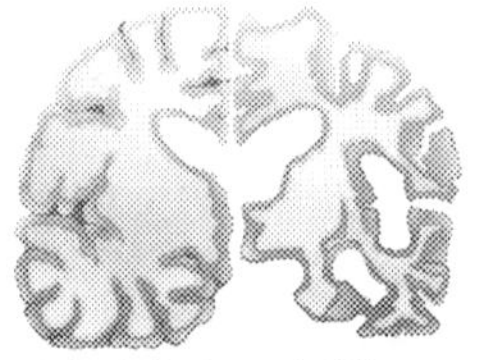

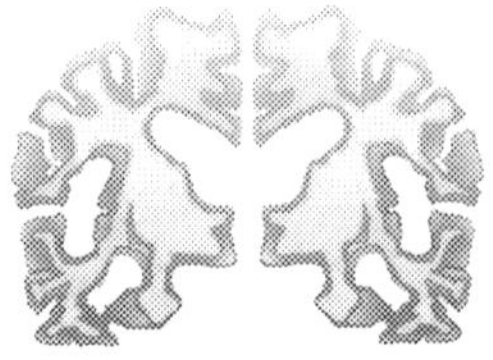

A hallmark of Alzheimer's disease is the progressive atrophy and shrinkage of the cortex of the brain. Scientists are working to regenerate some of the lost brain cells with stem cells.

even been found to reduce the plaques that choke the Alzheimer's brain, helping neurons regrow the insulating myelin sheaths that keep the brain's electrical signals from leaking out of the nerve cells.

Baumel and his team have also shown in mouse models that injections of mesenchymal stem cells encourage the growth of a structure in the brain called the hippocampus. This small, distinct area of the brain is almost exclusively where new memories are formed. New experiences and information are filtered through the hippocampus, which packages these memories for storage deeper in the brain. For this reason, neurogenesis—the growth of new neurons—in the hippocampus is a compelling goal for treating people with Alzheimer's, who can often recall their earliest and oldest memories, stored and protected deep in the brain, yet struggle to create, store, and recall new memories.

Again, this research is in an early stage, and the researchers don't want to oversell the promise of their technique. But, especially in a climate in which research funding is desperately tight, only the most promising clinical trials are allowed to move forward. The very fact that Baumel's trial is recruiting patients shows how far the treatment of Alzheimer's with stem cells has come. It has progressed from idea to basic biology with cells to mouse models of specific treatments and now, finally, based on many years of promising results, a human clinical trial of the technique. Millions of patients and concerned friends and family will be closely watching Baumel's trial. Is this the hope for Alzheimer's we have all been waiting for?

LOU GEHRIG'S DISEASE: THE NEW ROLE OF CELLULAR THERAPIES

Unlike a stroke, which is a vascular attack, neurodegenerative diseases destroy the mind and body slowly, their symptoms sometimes not visible to the naked eye for many years. Alzheimer's, for example, was "discovered" in a nursing home at the turn of the twentieth century. And very few people knew what amyotrophic lateral sclerosis (ALS) was until one of the most talented baseball players in the history of the sport was permanently sidelined by it in 1939.

When Lou Gehrig started the 1939 season with the New York Yankees, something was off. He was having trouble fielding fly balls and difficulty keeping his balance. It soon became clear that these mistakes were symptoms of a much larger problem. "I have seen ballplayers 'go' overnight, as Gehrig seems to have done," commented sports reporter James Kahn at the time. "But they were simply washed up as ballplayers. It's something deeper than that in this case, though."

On May 2, the "iron horse" of baseball told Yankees manager Joe McCarthy that he was benching himself "for the good of the team." Two years later, Gehrig, who had played 2,130 consecutive games, succumbed to ALS, giving his name to the disease and a face to the nearly 5,000 Americans who are diagnosed every year.

ALS is a progressive neurodegenerative disease in which the nerve cells in the spinal cord begin to deteriorate and die, causing the brain to lose its ability to initiate and control muscle movement. As the disease progresses, the patient becomes paralyzed. Death is inevitable. Most people with ALS die from respiratory failure, usually within three to five years of onset of symptoms. About half the patients, like Lou Gehrig, survive at least two years after diagnosis, while 20 percent live for

In 1939 baseball player Lou Gehrig brought global attention to amyotrophic lateral sclerosis (ALS), a neurodegenerative disease first described in 1869. (Wide World Photos Public Domain Photo)

five years or more. Only 10 percent of people with ALS survive more than a decade.

A resurgence of ALS awareness occurred in 2014 when videos of people dumping buckets of ice water on their heads in the name of ALS awareness swept the internet. The Ice Bucket Challenge was born of the video-sharing age and the entertaining idiocy of watching friends, family, and celebrities purposely douse themselves with freezing-cold water. The outlandish stunt paid handsomely, raising over $100 million dollars in donations to the ALS Association just in July and August of that year.

The Ice Bucket Challenge is a clear example of how money drives innovation. A $1 million grant from these donations directly funded a study called Project MinE, which supported researchers as they sifted through the data of 15,000 genomes of people with ALS and eventually found a gene that was linked to the disease. A 2016 letter in the journal *Nature* discusses the gene NEK1 and its relation to ALS. Efforts are underway to understand the gene's role in the disease, and it has been added to the list of possible new drug targets.

PROGRESS WITH STEM CELLS

Ice Bucket Challenge donations are also funding a parallel approach. Instead of unraveling the genetic causes of the disease and designing drugs to target dangerous genes or genetic changes, researchers are exploring the use of stem cells to fix the problem. ALS eats away at the body's motor neurons, nerve cells with long fibers that extend from the spinal cord out into muscles. Stem cells may be able to delay or prevent the decay of these cells.

At this writing, there are about thirty ALS-related stem cell trials underway, with another dozen recruiting patients. (Up-to-date information is available on the website ClinicalTrials.gov.) In most of the

trials, some form of stem cells, derived from fetuses or from the patients' own cells, are injected into the spinal cord or muscles of ALS patients. Emory University's Jonathan Glass, MD, for example, is leading a trial of fetal-derived neural stem cells. In the June 29, 2016, issue of the journal *Neurology*, Glass and his team reported that the procedure was safe for 13 of 15 patients who volunteered for the study.

Similarly, in early 2016 the *Journal of the American Medical Association (JAMA)* reported results of a study conducted in the United States and Israel that used stem cells derived from a patient's own bone marrow. The great news was that almost 90 percent of patients in this small study saw the progress of their disease slow. And in July 2016 the New Jersey company BrainStorm Cell Therapeutics reported that their experimental NurOwn therapy, using cells derived from a patient's own bone marrow, were not only safe, but also appeared to be more effective with those ALS cases that progress most quickly. All these trials are promising and with adequate funding should lead to larger and more definitive studies.

The most mature work with stem cells against ALS comes from the laboratory of Eva Feldman, MD, PhD, director of the University of Michigan Program for Neurology Research & Discovery, who has been using stem cells derived from adult tissues against ALS since 2010.

"One of the real heartbreaks of ALS is that as a patient progressively becomes weaker, the patient loses none of his or her abilities to think clearly, to understand what's happening to them, to really emote," Feldman said. Incapable of any communication whatsoever, ALS patients die prisoners in their own bodies. It is with this sense of compassion and urgency for sufferers of this disease that Feldman has completed phase I and phase II trials of adult-derived stem cells.

"If you take stem cells and place them into a diseased spinal cord, the large nerve cells that usually undergo degeneration with ALS begin to look healthier," Feldman says. Her description speaks to a new understanding of exactly *how* stem cells lead to healing in neurodegenerative conditions. The work of Feldman and others shows that stem cells may not create new tissue or become new neurons but instead may initiate a biochemical cascade that makes the brain and spinal cord more "plastic" and better able to repair itself.

Some researchers have likened this plastic state to an infant brain, in which connections, repairs, and rewiring happen much more fluidly than in the adult brain. You've heard about neuroplasticity and the brain's ability to rewire itself, pruning unneeded connections (called synapses) and generating new ones. Stem cells may "warm the plastic" of the brain, making it better able to grow, develop, and repair itself.

"Just replacing these cells wouldn't do the job," says Emory's Dr. Jonathan Glass, citing the fact that even pristine motor neurons would lack the necessary connections with the brain, with each other, and with the muscle tissues they are supposed to control. Still, he is hopeful these stem cells will help to save existing motor neurons from the decay associated with ALS.

Stem cells may also signal the need for repair, naturally seeking out damaged areas of the brain or, in the case of ALS, motor neurons of the spinal cord. In this scenario, stem cells would signal the release of various neuro-protective and neuro-reparative chemicals.

At the 2015 meeting of the American Neurological Association, Feldman, the association's newly elected president, presented the results of her phase II clinical trial. Since ALS is a degenerative disease, all 15 patients in her trial would be expected to lose function progressively. Unfortunately, this is exactly what happened, and yet 70 percent of patients lost function *more slowly* than would have been expected. This trial wasn't meant to demonstrate definitive results, but it presents compelling evidence that stem cell therapy can slow the advancement of ALS.

Of course, stem cell treatments for ALS already exist outside the United States. At clinics in China, India, Mexico, and elsewhere, patients with $50,000 can purchase the injection of stem cells into their spinal cords, at their own risk.

"Unfortunately, there are snake oil salesmen out there taking advantage of vulnerable people," says Ericka Simpson, MD, a nationally recognized ALS researcher at Houston Methodist Hospital. However, the devil—or in this case the angel—is in the details, and while Simpson sees the danger of unproven and unregulated stem cell procedures, she also sees the promise of controlled and carefully designed treatments.

In an article for the Dana Foundation, Glass writes, "It is an exciting time, but we must move forward with meticulous regard for the scientific process and cautious respect for what we do not know and cannot anticipate."

PARKINSON'S DISEASE AND STEM CELL THERAPY

Parkinson's disease is a degenerative neurological movement disorder that affects about 1.5 million people in the United States. It is named after James Parkinson, MRCS, the British physician who in 1817 was the first to accurately describe the disease. Symptoms tend to start with a rhythmic shaking (tremor) and progress to slowed movements and problems with gait. Some patients also eventually suffer from a form of dementia that includes confusion, memory loss, and delusions. About one of every 100 people over the age of sixty in the United States has Parkinson's. Although the disease is primarily associated with older people, many patients are diagnosed before age fifty.

Parkinson's is caused by the breakdown or death of the neurons in the brain that produce dopamine, a substance necessary for smooth and coordinated movements. Scientists have dreamed for years of replacing dopamine cells as a treatment strategy. Thanks to the stem cell work of Ole Isacson, MD, founder and former director of the Neuroregeneration Research Institute at McLean Hospital in Belmont, Massachusetts, and now the chief scientific officer of the neuroscience research unit at drug giant Pfizer, we are getting closer to that reality. Dr. Isacson and his colleagues at McLean reported in the journal *Cell Reports* that after successfully transplanting dopamine-producing cells

At age eighteen Muhammad Ali won a gold medal in boxing at the 1960 Summer Olympics in Rome. By age twenty-two he was the heavyweight boxing champion of the world. After conscientiously objecting to the Vietnam War, he was controversially convicted for draft evasion and stripped of his title. That judgment was later overturned by the Supreme Court, and Ali regained his heavyweight title. He was diagnosed with Parkinson's Syndrome in 1984, which gradually took his mobility and ability to speak. He died in 2016. (Library of Congress, Prints & Photographs Division, NYWT&S Collection [reproduction number, e.g., LC-USZ62-115435])

into the midbrains of adult patients with late-stage Parkinson's, the cells remained healthy and functional for up to fourteen years.

These findings are critically important for the development of stem cell therapy for Parkinson's. Historically, there has been skepticism that transplanted dopamine cells could remain healthy without themselves succumbing to the pathology of Parkinson's. With this pivotal study, however, Isacson and his team disproved that notion. About one year after transplantation, the cells finally matured and took hold, at which point the patients who received the neuron boost no longer needed their dopamine replacement medication. "Patients with Parkinson's usually get worse by 10 percent a year," says Isacson. "If these study patients had continued without the cell transplants, they would have eventually been in wheelchairs. Instead, they were markedly improved and looked like young Parkinson's patients with some deficits—and they were off drugs."

While study participants received cells harvested from human fetuses at its start over fifteen years ago, both ethical issues and the long, difficult process of extraction have led to the development of new research possibilities. Isacson is now working to develop dopamine neurons from iPSCs that are made from a patient's own stem cells and grown in the lab.

There is no drug that can stop the devastating progression of Parkinson's disease, but many experts think that with stem cells, there could be. Australian researchers recently launched a phase I clinical trial using stem cells derived from female patients' unfertilized eggs, a strategy developed by the International Stem Cell Corporation, a California-based biotechnology company. Since these eggs don't have the potential to become embryos, the ethical dilemma surrounding the use of embryonic stem cells has been avoided.

The clinical study, which was launched in 2016 at the Royal Melbourne Hospital in Australia, is a "dose escalation safety and preliminary efficacy study" of the stem cells, in which thirty to seventy million cells were transplanted into areas of patients' brains affected by the disease. The first patient, a sixty-four-year-old man, was treated by neurosurgeon Girish Nair, MBBS, MCh. In a five-hour procedure, Nair inserted the cells through two holes in the man's skull, targeting fourteen specific sites on the brain, seven on each side. The study will use PET scans and various Parkinson's disease rating scales to evaluate the safety and biologic activity of the stem cells and to watch for improvements in tremor, rigidity, walking, and ability to express emotions up to a year after treatment. Treatment of the remaining study patients will be completed in 2017 with results to be reported in 2019.

OVER THE HORIZON

Again, there are no real cures yet for the deficits of chronic stroke, Alzheimer's, ALS, Parkinson's, or many of the other neurodegenerative diseases that affect people around the world. But increasingly and with more momentum than ever before, researchers, doctors, and the brave patients like Sonia Olea Coontz who agree to test these treatments are demonstrating—and sometimes proving—that human cells are the medicines we have been searching for all along.

With care and creativity, we are finally moving into an era in which cells will be the drugs used to slow, stop, or even reverse the ravages of diseases of the brain.

CHAPTER FOUR

STEM CELLS AND CANCER

More than 10 million Americans are living with cancer, and they demonstrate the ever-increasing possibility of living beyond cancer.

—SHERYL CROW

The World Health Organization estimates that in the next two decades the global cancer burden will rise to 22 million new cases annually and 13 million deaths per year. In 2012 these numbers were 14 million and 8.2 million, respectively.

Cell therapy is among the most exciting and promising fields of research for cancer treatments. But you might be surprised to learn that using stem cells as part of cancer therapy isn't new. As a matter of fact, the first attempts to fight cancer with stem cells were made in the 1950s. This work began as a radical approach to treating leukemia, cancers of the blood, pioneered by E. Donnall Thomas, MD, whose hypothesis was that if a patient's blood system was diseased, then removing and replacing it, rather like changing the oil in a car, should fix the problem.

By 1957 Dr. Thomas had performed six bone marrow transplants on humans. He used radiation to kill the patient's blood system and then introduced new blood stem cells from a healthy donor's bone marrow.

Unfortunately, all six transplants failed. Either the patient's immune system would attack and kill the new bone marrow stem cells or else the new stem cells would ravage the patient's body, eating away at lung and heart tissue. Due to what is called graft-versus-host disease (GVHD), none of Thomas' first 6 patients lived more than 100 days.

What was it about blood components that might make them medicine for one person and poison for another? The answer turned out to be human leukocyte antigen (HLA) proteins, the many kinds of cell-surface markers that define whether cells are seen as "self" or "foreign" and therefore whether a patient's immune system will attack a donor's blood or be attacked by it. As Thomas found, HLA proteins are the difference between a chance at life and almost certain death for transplant patients.

In 1968 the first successful bone marrow transplants against leukemia were performed at an underground military bunker in West Seattle. In order to destroy their cancer-ridden bone marrow, Thomas' patients were irradiated with beams from a cobalt-60 source that had been designed to study the effects of an atomic bomb blast. They were then rushed from the bunker to a sterile transplant ward at the Seattle Public Health Service Hospital.

Of 54 patients treated this way, 6 were cured of their leukemia. In 1979, Thomas reported that cure rates had risen to 50 percent. Today, depending on the type of leukemia and the age and fitness of the patient, cure rates from bone marrow transplant exceed 70 percent and are higher than 90 percent for some cancers.

Thomas' work earned him the 1990 Nobel Prize in Physiology or Medicine. It also demonstrated the power of stem cells to treat cancer.

Dr. E. Donnall Thomas (left) accepting the Nobel Prize for his pioneering work in bone marrow transplantation. (Photo by The Nobel Foundation; photo courtesy of Fred Hutchinson Cancer Research Center)

Today, researchers are exploring new uses for stem cells against cancer. But before discussing those, let's look at a strategy that uses stem cells to fine-tune Thomas' original technique of bone marrow transplant.

TREATING GVHD WITH STEM CELLS

When radiation, chemotherapy, and targeted treatments fail to control a patient's blood cancer, the next step is often bone marrow transplant. According to the Fred Hutchinson Cancer Research Center, a world leader in the treatment of blood cancers and Thomas' home base after its founding in 1975, about 95 percent of leukemia patients will be able to find a donor who matches their all-important HLA proteins. The match will come either from their immediate family (30 percent) or from the BeTheMatch.org registry. Because HLA proteins are linked to ethnicity, people such as Native Americans, African Americans, and Latinos may have a lower chance of finding a registry match.

Doctors test for matches in six to ten different HLA proteins. Ideally, a donor's blood will be, for example, a 6/6 match with the patient's blood. However, a 5/6 or even 4/6 match can make for a successful transplant. In fact, some conflict can be good. It's often impossible to completely irradiate a patient's blood system, which would leave the patient completely at the mercy of even the slightest infection. Therefore, when a donor's stem cells are infused, remnants of the patient's original, cancerous blood system often remain. A slight immunological mismatch helps to ensure that any last trace of old blood is destroyed.

Unfortunately, a donor's immune system may not know when to apply the brakes. Instead of the body rejecting the new blood system, the transplant can, in essence, try to reject the body into which it has been placed. Weeks or months after the procedure, some patients develop the frightening signs of the GVHD that attacked Thomas' first transplant patients in the 1950s. In GVHD, rogue immune cells from the donor's blood system attack the skin, digestive system, liver, lungs, connective tissues, eyes, and other tissues.

According to the Center for International Blood and Marrow Transplant Research (CIBMTR), there are about 30,000 bone marrow transplants every year, and about half of these patients will go on to develop some form of GVHD—about 5,500 people per year will develop acute, life-threatening GVHD.

In 2008 the journal *Oncology* described the plight of a patient with GVHD. A patient *Oncology* identified as Mr. SR, age thirty-eight, had received a bone marrow transplant from his sister to treat the blood cancer non-Hodgkin's lymphoma. Mr. SR received immunosuppressant drugs, and doctors prescribed a maintenance dose of the steroid prednisone to support his body's efforts to resist. It didn't matter—he still developed GVHD.

During the following year, the degree and intensity of his GVHD varied, and it was difficult to manage. He recovered from one flare-up only to have another. He developed a pus-leaking rash across his back, hands, and legs. Eventually, he told his doctors that he "felt like a freak" and that his family was "grossed out." His wife asked for a divorce. He had been forced to quit his job and was living with his uncle. Mr. SR had survived non-Hodgkin's lymphoma, but GVHD had ruined his life.

Imagine how devastating it is to be cured of one life-threatening disease only to learn that the cure itself had caused another. In severe cases of GVHD, the time from diagnosis to death can be about 80 days. When GVHD attacks the liver, the risk of death can be as high as 85 percent.

As was the case with Mr. SR, steroids may help a patient resist GVHD, and immunosuppressant drugs may tamp down the condition's symptoms, but there are no therapies specifically approved to treat the underlying cause of GVHD—at least not in the United States.

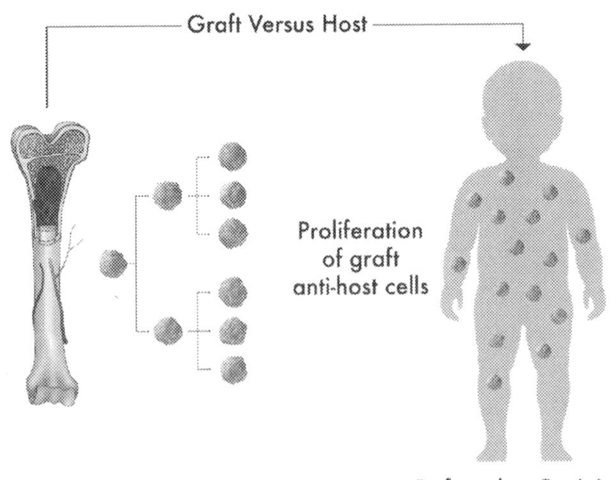

Graft-versus-host disease (GVHD) is a potentially life-threatening complication of a bone marrow or stem cell transplant. It is the result of the transplanted cells attacking the recipient of the graft.

In Japan, patients are now using an "off-the-shelf" (i.e., already processed, not fresh from a donor or the patient's own) mixture of adult mesenchymal stem cells to treat GVHD. The cell-based product TEMCELL® HS Inj. (developed by JCR Pharmaceuticals Co., Ltd., licensed in from Mesoblast) is made from mesenchymal stem cells extracted from a healthy person's bone marrow. As we've seen, mesenchymal stem cells are useful in repairing damaged tissue, but they can also be used to modulate the immune response in GVHD. Importantly, these stem cells don't express the HLA proteins that could trigger a patient's immune system, meaning that these cells can be given to any patient without fear of rejection.

In Sweden, the Netherlands, Italy, and Australia, 55 patients with steroid-resistant acute GVHD were infused with mesenchymal stem cells from donors who were randomly chosen, not matched. Sixty percent of these patients had failed at least two rounds of immunosuppression therapy. In other words, their prognoses were not good. A study in *The Lancet* noted that "just over half the patients had a complete response." Even more promising, of the 25 children who were part of the trial, 17 were completely cured of their GVHD. Overall, 39 of the 55 patients showed some response to the treatment.

"Don't think of MSC-100-IV, Mesoblast's cell therapy for steroid refractory acute graft versus host disease, as one drug, but as many, all responding to signs from diseased tissues," says Donna Skerrett, MD, chief medical officer of Mesoblast. What she means is that where most drugs have only one purpose and one effect, medicines like MSC-100-IV are multifactorial, offering the many beneficial effects of mesenchymal stem cells that have evolved to aid healing.

Indeed, the *Lancet* article states, "At present, little is known about mechanisms of suppression of GVHD by mesenchymal stem cells." It may be that stem cells disarm the new immune system's attack on patients' tissues, making donors' T cells less aggressive. It may be that stem cells create conditions of healing around damaged tissues that allow them to survive the onslaught. Or it may be that mesenchymal stem cells are supporting patients via some mechanism we haven't imagined yet. What we do know is that an ever-increasing wave of evidence suggests that infusions of third-party mesenchymal stem cells have the real potential to treat and even cure GVHD.

At the 2016 scientific meeting of the American Society for Blood and Marrow Transplantation (ASBMT), Duke University's Joanne Kurtzberg, MD, announced results of a phase III trial evaluating MSC-100-IV in 241 children with steroid refractory acute GVHD. Overall, 65 percent of the children responded to treatment. Moreover, only 39 percent of the children who were not treated with MSC-100-IV survived 100 days postdiagnosis while 82 percent of the children who were treated with MSC-100-IV were alive at 100 days postdiagnosis and treatment. This clinical trial *doubled* the chance that kids would survive more than 100 days after their treatment for often-fatal GVHD.

"There is a critical and urgent need for an effective and well-tolerated treatment for the very ill children who develop this life-threatening complication after a bone marrow transplant," says Kurtzberg, who is the Jerome S. Harris Professor of Pediatrics at the Duke University School of Medicine. "While historically there is a high mortality rate associated with this complication, we are now seeing the majority of children who receive Mesoblast's cell therapy respond and survive."

It can be difficult to earn FDA approval for a drug that shows promise without knowing how the drug works. It can be even more challenging when the drug is not a molecule synthesized in the lab but living cells that may go on living and replicating in the patient's body. And approval can be *triply* challenging for stem cells.

However, while we wait, children are dying. Without successful treatments for GVHD, patients who survive cancer, including children, will die from complications associated with their treatment. Now, with strong clinical trial results and with the potential approval of Mesoblast's cell therapy MSC-100-IV, we are very close to adding this exciting cell-based therapy to the list of approved medicines that save lives.

STEM CELLS AS DRUG DELIVERY

A major hurdle in humanity's plans to colonize Mars is what you can't take with you. The tools and supplies colonists would need on the planet are simply too big and too many to fit in any imaginable spacecraft. Instead, NASA is exploring ways to send small ships to Mars, each with bits and pieces of the things colonists would need, which could then be assembled once they arrive. This outer-space strategy is promising for inner space, too. Specifically, where it is difficult or impossible to deliver cancer medications to the brain, scientists may be able to break apart medicines and use stem cell spaceships to deliver the pieces.

The obstacle in question is the blood-brain barrier. This membrane is your body's last line of defense around your most fragile and important organ. It admits only the oxygen and key nutrients your brain needs to power itself, which keeps your brain and central nervous system safe from infection. The problem is that most drugs that doctors use to treat cancers, infections, or other neurological conditions are far too large to slip through the blood-brain barrier.

This is why the brain cancer known as glioblastoma multiforme is so deadly. Many cancers spread to the brain, but glioblastoma multiforme starts there. And instead of forming a tightly packaged centralized tumor that a surgeon can cut out, these tumors tend to spread invasively, which means they send thin tendrils or diffuse waves through the brain that no scalpel can reach.

Elsewhere in the body, doctors are able to treat invasive cancers with a range of targeted therapies and chemotherapies, but, as noted, many of these medicines are too large to pass through the blood-brain barrier. (One exception is the drug temozolomide, which somewhat successfully targets cancer cells' ability to replicate rapidly.) Also, while it may be acceptable to kill some healthy tissue surrounding other cancers in the body, collateral damage in the brain has dire consequences for such basic brain functions as thinking and movement. The chemotherapies that *are* able to enter the brain kill too indiscriminately to be usable.

The combination of these two constraints—molecule size and the need to kill cancer cells without damaging surrounding tissue—means that most glioblastomas are not treatable. This aggressive brain tumor affects about 10,000 people every year, and fewer than 5 percent will live five years from diagnosis.

Stem cells may provide the solution to the brain-barrier blockade. Stem cells are small enough to pass through the blood-brain barrier, and they naturally seek damaged areas of the body, including those disturbed by cancer.

When researchers introduce stem cells into animal models, they can see these cells congregate around tumors. Questions remain about how to make use of this ability. For example, should doctors use donated "off-the-shelf" mesenchymal stem cells, or are neural stem cells best equipped to seek tumor tissue in the brain? And how many stem cells and by what method should they be introduced to patients? We have the potential vehicle that can specifically seek brain tumors, but what should the payload be?

There are a couple of candidates. One idea is to attach tiny molecules of chemotherapy directly to stem cell "vehicles." There are two problems with this strategy. First, some chemotherapies kill the stem cells to which they are attached. Second, any toxic chemotherapy in the brain may be too much. Nonetheless, work on developing this strategy continues. Another idea is to use stem cells to deliver gene therapy, perhaps by delivering small molecules of RNA that delete essential genes from cancer cells, or, in the future, to use gene-editing technologies such as the CRISPR system to edit cancer cell genomes directly (more on this in chapter nine). But gene therapies bring their own concerns about safety and efficacy.

One approach that seems most futuristic, at least on the surface, is, in fact, closest to being used widely. At City of Hope in Duarte, California, Karen Aboody, MD, recently completed a 15-person clinical trial that involved bringing together stem cells "building supplies" at the site of a tumor, as in our mission-to-Mars analogy.

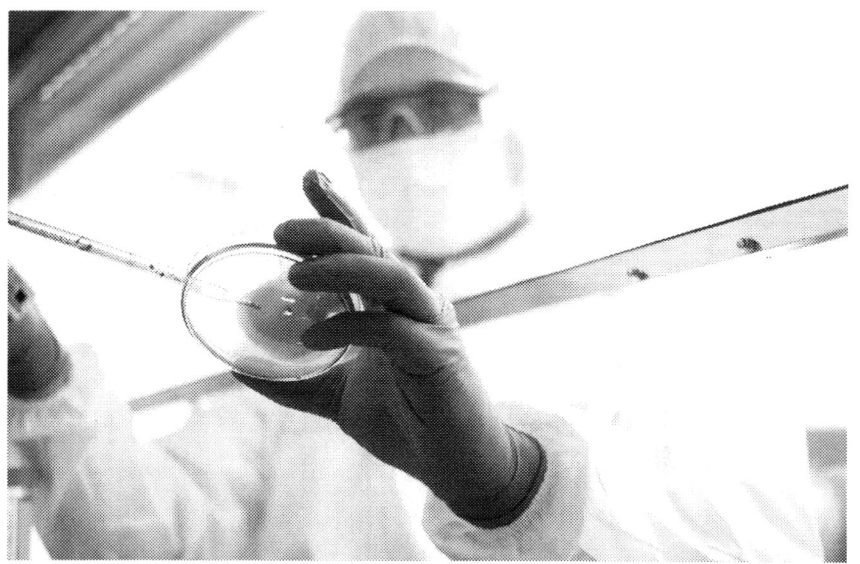

The trial required considerable engineering. First, Aboody and her team tweaked the genomes of neural stem cells to make a new protein called CD. By itself, CD was harmless. She then engineered the prodrug 5-fluorocystosine which, by itself, was also harmless. (A prodrug is a medication that only has an effect after it has been metabolized.) When CD came in contact with 5-FC, the result was the manufacture of the chemotherapy drug 5-fluorouracil, otherwise referred to as 5-FU. Interestingly, 5-fluorouracil was one of the first chemotherapies, described in a 1957 paper in the journal *Nature*. Now, these building blocks of engineered neural stem cells and the prodrug 5-FC have the potential to bring this historical chemotherapy to the modern treatment of brain cancer.

Aboody's 15-patient trial was an early-phase experiment, with the primary goal of showing that the treatment was safe. It was. The secondary goal was to show that stem cells could combine with the prodrug at tumor sites to make the chemotherapy. It worked. Now Aboody and her team plan a more formal phase I trial of their treatment.

Treatments based on mesenchymal stem cells aren't far behind. In fact, certain aspects of studies that use these adult-derived cells are showing results at least as impressive as those using neural stem cells. For example, a study in the *International Journal of Cancer* used stem cells derived from fat tissue to cure glioblastomas in rodent models. The researchers used these cells to see if they could make the accepted treatment better. They surgically removed tumors from rat brains and then, while they were still inside the blood-brain barrier, introduced special mesenchymal cells into half the group with diffuse waves of cancer cells that spread and could not be resected.

What made these mesenchymal cells special is that they were engineered with a "suicide gene" that killed the cells once they had served their purpose as medicine, so as to eliminate the possibility of these cells damaging the body after the therapy.

As in Aboody's trial with neural stem cells, this trial used mesenchymal stem cells to deliver a prodrug that manufactured 5-FU. "In this therapeutic arrangement, we observed a strong inhibition of tumor growth leading to curative therapy in a significant number of animals," the authors wrote.

Now you know about many of the things that stem cells can do. They can grow new tissue, induce growth factors that stimulate the protection and healing of tissues, and regulate the immune system. It seems that stem cells, even those derived painlessly from adult tissues, can deliver essential drugs directly to deadly brain tumors to destroy cancer cells while preserving good tissue. Stem cells are really quite extraordinary. Every day, around the world, researchers learn more about them—their ability to work like drugs, to help deliver drugs, and even to potentiate the way drugs work.

Scientists and institutions have initiated several major programs to accelerate finding cures for cancer using stem cells as well as other cells of our immune system, including NK cells and T cells, to enhance the immune system's ability to fight cancer cells. The odds are very high that some of these therapies will soon be saving the lives of people with inoperable tumors and metastatic cancers.

PART TWO

OUR IMMUNE SYSTEM AS WEAPON AND HEALER

Natural forces within us are the true healers of disease.

—HIPPOCRATES

CHAPTER FIVE

TEACHING THE BODY TO FIGHT CANCER

When you have exhausted all possibilities, remember this: You haven't.

—THOMAS EDISON

When former president Jimmy Carter was diagnosed with melanoma that had spread from his liver to his brain, the country cringed at the death sentence. Still, we told ourselves, he had lived a long life. He had spent his last years giving back. Everyone has to go sometime. Yet the months ticked by, and Jimmy Carter didn't go. In September 2016, more than a year after his diagnosis, Carter and his wife, Rosalynn, were in Memphis, Tennessee, swinging hammers to frame a house with the charity Habitat for Humanity. Scans showed that his cancer was nearly undetectable.

You have heard miraculous stories of drugs that kill cancer. And while Jimmy Carter's story is certainly that of an astounding new

President Jimmy Carter (1977–1981) was diagnosed with metastatic melanoma in the brain and the liver at age ninety. (Courtesy of the Jimmy Carter Presidential Library and Museum)

medicine, his drug doesn't actually kill cancer at all. Jimmy Carter's treatment, called pembrolizumab (KEYTRUDA), shines a spotlight on cancer so that the cells of the immune system can do the job they are designed for—attacking dangerous cells.

Immune therapy for cancer was born one day in 1891 when William B. Coley, MD, famed surgeon at Memorial Hospital in New York, was reviewing the patient record of housepainter and German immigrant Fred Stein. Seven years earlier, Stein had had surgery to remove a fast-growing tumor from his neck. The surgery was unsuccessful and the tumor continued to grow—until Stein happened to develop a bacterial skin infection called erysipelas on his face. To everyone's surprise, soon after the infection set in, Stein made a full recovery from his cancer.

For Coley, this was an aha moment. Had the infection awakened Stein's immune system, causing it to defeat the cancer? Coley reviewed the medical literature and found more than four dozen other cases in which infections had improved cancer, and with each case, Coley became more and more convinced that humans come equipped with the tools needed to fight cancer.

In 1893, perhaps benefiting from that era's lax controls on the use of human subjects, Coley injected twenty-one-year-old sarcoma patient John Ficken with a bacterial cocktail labeled "Coley's toxins." Coley's hope, based on what today would be considered only the most rudimentary understanding of the biology, was that he could provoke an immune response that would be directed not only at the bacteria but also at his patient's cancer. Astoundingly, and with no small measure of serendipity, Coley's therapy worked. Ficken's cancer remained in remission until his death from a heart attack twenty-six years later.

Coley reported the results of this and nine other uses of Coley's toxins against cancer in an 1893 issue of the *American Journal of the Medical Sciences*. He went on to treat more than a thousand patients with versions of bacterial "toxins," finding at least some success. Unfortunately, as radiation and chemotherapy became standard treatments for cancer, Coley and his toxins were pushed aside. His patient successes were contested, and due to behind-the-scenes politics at Memorial,

Dr. William B. Coley a pioneer in cancer immunotherapy through his work using "Coley's toxins." (Cancer Research Institute, http://cancerresearch.org)

brokered deals with philanthropists, and claims of poor medical record keeping, Coley was eventually forbidden to administer any of his cancer therapeutics even at his own hospital.

CANCER IMMUNOTHERAPY COMES OF AGE

Despite dramatic advances in our understanding of cancer and of the immune system, immuno-oncology was still considered as recently as a decade ago to be a fringe science, with cancer researchers split in their opinions of whether or not it would ever pan out.

One major problem is that your immune system is exquisitely balanced. Like a coin resting on its edge, it's easy to push the immune system in either direction. If the coin falls to the side of less sensitivity, it misses bacteria, viruses, and other invaders that would harm the body. If it falls to the side of more sensitivity, it attacks the body's own tissue, resulting in autoimmune conditions like rheumatoid arthritis, type 1 diabetes, and lupus.

No one had found a way to dial back the immune system without causing it to either miss or kill things that you would rather not miss or kill. The challenge is to turn up the immune system selectively, teaching it to be more sensitive against cancer but not against the rest of your body.

From World War II through the 1980s, the promise of effective cancer immunotherapy hung tantalizingly just out of reach of doctors and scientists. In the meantime, researchers came to understand the mechanisms of immune surveillance, the process by which the immune system stays on the lookout for antigens presented by invading and foreign cells. They came to understand how the immune system's B cells develop antibodies matching these antigens, enabling B cells to neutralize or destroy recognized invaders. What many people don't realize is that the immune system also plays a vital role in protecting us against cancer.

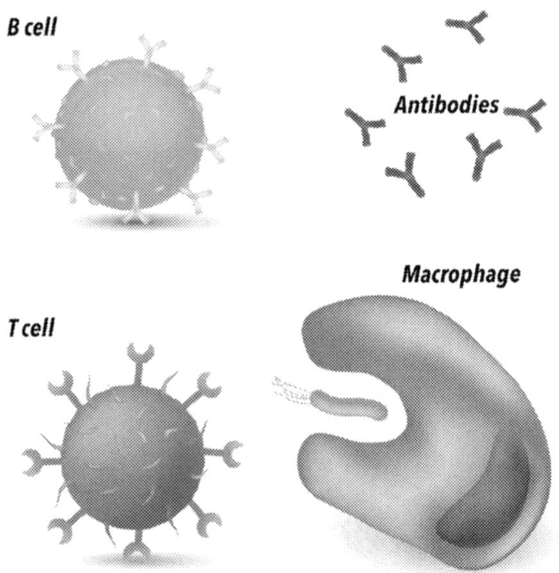

Components of the human immune system.

With tens of billions of cell divisions happening in our bodies every day, it's inevitable that some errors in DNA duplication will occur. Most of those mutations will be either harmless or so detrimental to cellular function that the cell simply dies. But occasionally a mutant cell does survive, one that could become cancerous. That's where the immune system steps in, recognizing the cell as harmful and destroying it. It's likely that our bodies make these potentially cancerous cells daily, while immune surveillance constantly protects us, keeping cancer from becoming an everyday occurrence.

But very rarely, one of these mutant cells is able to both survive and evade detection by means of the cellular version of Harry Potter's cloak of invisibility. These mutant cells are "invisible" in part because cancer grows from the body's own tissues, which means that the immune system may have trouble recognizing cancer cells as different from healthy tissue. The basis for the newly exploding field of cancer immunotherapy is finding one or more ways to strip a cancer cell of its invisibility or to mark cancer cells in ways that make them more visible.

War on cancer: President Richard M. Nixon signs into law the National Cancer Act of 1971. (Courtesy of the National Cancer Institute)

For the last half century or so, science has been engaged in a "war on cancer," the results of which have been mixed. The world marveled at our astonishing progress in chemotherapies and surgical techniques and our ability to target cancer with radiation. But far more valuable in the long run was the far less sexy knowledge scientists were gathering about the basic biology of cancer cells, their genetics, biochemical pathways, and the insidious ways in which they spread and evade the immune system.

For example, the cancer war has helped us understand the genetic underpinnings of the many different types of cancer. No longer is cancer seen as one disease—not even all breast cancers are alike. We now understand that cancer is a family of related diseases, each with distinct genetic causes and, potentially, distinct cures.

In gaining this knowledge, oncologists realized that cancers should be classified by their genetic and biochemical characteristics, rather than by the organ or location in which the tumor happens to grow.

Some breast cancers, for example, more closely resemble some prostate cancers than other breast cancers.

DR. JIM ALLISON AND THE MODERN ERA OF CANCER IMMUNOTHERAPY

As it became clear that cancer is not a single disease, hope for a single, magic-bullet cure faded. Instead, researchers started looking for specific cures for specific cancers, and much of that research has been focused on biomarkers. By comparing the genetics of cancer to the genetics of healthy tissue, researchers started finding the biomarkers that made cancer cells different, and they began choosing treatments based on cancer's defining genetic features. In 1987 French researchers identified a new protein on the surface of T cells, the type of white blood cells that recognize and attack invaders. The protein was cytotoxic T-lymphocyte-associated protein 4 (CTLA-4).

Until 1996 no one knew what it did. In that year, however, James P. Allison, PhD, a Texas-born harmonica-playing Willie Nelson devotee working in a lab at the University of California at Berkeley, showed that CTLA-4 neatly and effectively blunts the activity of T cells.

"The immune system has natural 'brakes' that are used to stop immune responses at the right time," says Allison. "When many cancers develop, they also take over this braking mechanism, which allows the cancer to become invisible to the immune system."

Allison is now professor and chair of immunology at M.D. Anderson Cancer Center and director of the Cancer Research Institute (CRI) scientific advisory council. The CRI, a group founded in 1953 by Helen Coley Nauts, the daughter of Dr. William B. Coley, is the world's leading nonprofit organization dedicated exclusively to harnessing the immune system's power to conquer all cancers.

Allison's groundbreaking immunotherapy work, which eventually earned him the 2015 Lasker Prize, sometimes referred to as the "US

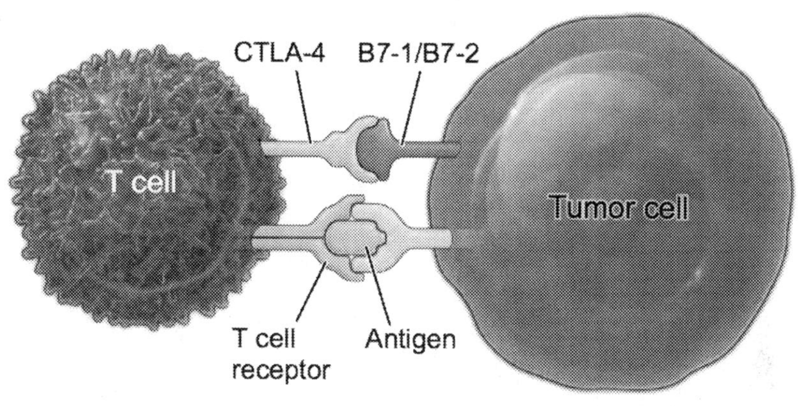

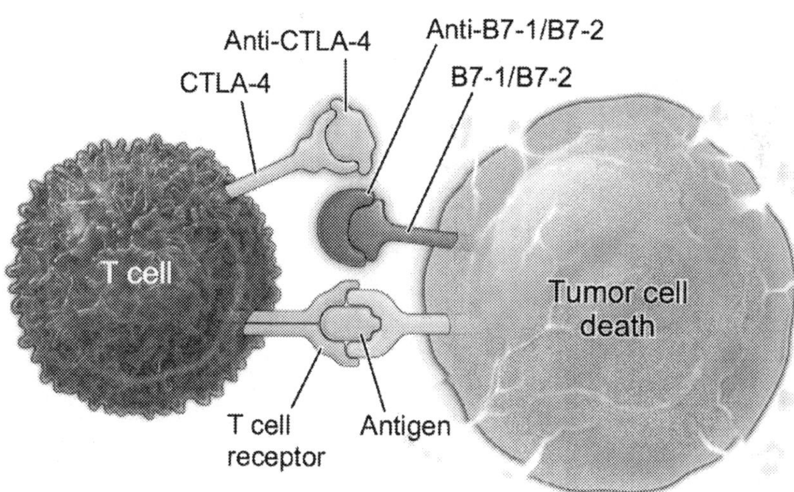

T cells use special receptors to attach to antigens on cancer cells (top). But tumor cells have other surface molecules (B7-1/B7-2) that also attach to different T-cell receptors (CTLA-4), rendering the tumor "invisible" to the immune system, which would normally kill it.

The lower diagram shows how an anti-CTLA-4 drug such as ipilimumab or an anti-B7-1/B7-2 drug can release the brakes on T cells and allow them to kill tumor cells.

The Lasker Awards have been awarded annually since 1945 to living persons who have made major contributions to medical science or who have performed public service in behalf of medicine. The awards, which are sometimes referred to as the "American Nobel Prizes," are administered by the Lasker Foundation, founded by Albert Lasker and his wife, Mary Woodard Lasker.

2012

Albert Lasker Basic Medical Research Award
Motor proteins that contract muscles and enable cell movements

Michael Sheetz, PhD
James Spudich, PhD
Ronald Vale, PhD

Lasker-DeBakey Clinical Medical Research Award
Liver transplantation

Sir Roy Calne, FRS
Thomas E. Starzl, MD, PhD

Lasker-Koshland Special Achievement Award in Medical Science
Fundamental biomolecular techniques

Donald D. Brown, MD
Tom Maniatis, PhD

2013

Albert Lasker Basic Medical Research Award
Regulated neurotransmitter release

Richard H. Scheller, PhD
Thomas C. Südhof, MD, PhD

Lasker-DeBakey Clinical Medical Research Award
Modern cochlear implant

Graeme M. Clark, MBBS, MSurgery, PhD Med, MD
Ingeborg Hochmair, PhD
Blake S. Wilson, PhD

Lasker-Bloomberg Public Service Award
Advancing global health through enlightened philanthropy

Bill and Melinda Gates

2014

Albert Lasker Basic Medical Research Award
Unfolded protein response

Kazutoshi Mori, PhD
Peter Walter, PhD

Lasker-DeBakey Clinical Medical Research Award
Deep brain stimulation for Parkinson's disease

Alim Louis Benabid, MD, PhD
Mahlon R. DeLong, MD

Lasker-Koshland Special Achievement Award in Medical Science
Breast cancer genetics and human rights

Mary-Claire King, PhD

2015

Albert Lasker Basic Medical Research Award
Discoveries concerning the DNA-damage response

Stephen J. Elledge, PhD
Evelyn M. Witkin, PhD

Lasker-DeBakey Clinical Medical Research Award
Unleashing the immune system to combat cancer

James P. Allison, PhD

Lasker-Bloomberg Public Service Award
Sustained and effective frontline responses to health emergencies

Médecins Sans Frontières

2016

Albert Lasker Basic Medical Research Award
Oxygen sensing—an essential process for survival

William G. Kaelin Jr., MD
Sir Peter J. Ratcliffe, FRS
Gregg L. Semenza, MD, PhD

Lasker-DeBakey Clinical Medical Research Award
Hepatitis C replicon system and drug development

Ralf F.W. Bartenschlager, PhD
Charles M. Rice, PhD
Michael J. Sofia, PhD

Lasker-Koshland Special Achievement Award in Medical Science
Discoveries in DNA replication, and leadership in science and education

Bruce M. Alberts, PhD

Albert and Mary Lasker Foundation

Cancer researcher Dr. James Allison. (Courtesy of The University of Texas MD Anderson Cancer Center)

Nobel Prize," was in the development of an anti-CTLA-4 drug that could remove this braking function from the immune system.

Sharon Belvin knows about Dr. Jim Allison's remarkable work firsthand. On May 28, 2004, Sharon was twenty-two years old, had just graduated from West Virginia University, and was busy planning for her June wedding. Was the stress of change the reason she suddenly seemed to be constantly out of breath? A longtime recreational runner who regularly logged five-mile workouts, she had been finding it difficult to run and talk at the same time. Walking up a flight of stairs was suddenly a major feat.

Finally, on that day in May, a biopsy and CAT scan revealed exactly what Sharon had feared—cancer. She had advanced melanoma, the deadliest form of skin cancer. And it had already spread throughout her body, including her lungs. At best, her oncologist gave her a 50 percent chance of surviving the next five months.

As expected, none of the traditional therapies worked very well or for very long. Looking beyond the horizon of established medicines,

Melanoma researcher Dr. Jedd Wolchok. (Courtesy of Memorial Sloan Kettering Cancer Center)

Sharon consulted with Jedd Wolchok, MD, PhD, a leading melanoma specialist at Memorial Sloan Kettering Cancer Center in New York City. Wolchok was a colleague of James Allison, then also working at the center. He explained to Sharon about a new clinical trial of a drug Allison had developed. The drug, called ipilimumab, or "ippy" for short, was meant to block the action of the immune system's brake, CTLA-4.

Ippy hadn't been tested in humans but had done well in mice, which couldn't have inspired a great deal of confidence. Still, Sharon had few other options, so she agreed to give it a try.

"She was so sick at that point," recalls Wolchok, "that she was really on the borderline of being able to enroll in our drug study. She eventually did land a spot in the study, however, and began treatment immediately."

The ippy was infused into one of Sharon's veins in a ninety-minute outpatient procedure that was repeated every three weeks. "After four treatments," says Wolchok, "we scanned her body to see what had happened. The tumors in her lungs had shrunk significantly, thanks to the drug. Sharon had a complete response."

A year later, back at Wolchok's office for a follow-up visit, Sharon got the news that she never expected to hear. "You no longer have cancer," Wolchok told her. Understanding that the ipilimumab treatment was a team effort, Wolchok asked Sharon if she wanted to meet James Allison, the researcher behind CTLA-4 and the drug that had saved her life.

"I really knew nothing about Sharon," says Allison. "I was working in my lab when Jedd called, urging me to come to his office right away to meet some special people. Of course, I was initially reluctant. I never saw patients, and besides, my lab was almost twenty blocks away, which was quite a hike. But Jedd persisted, and so I immediately set out.

"When I finally got there, Sharon was there with her father. She came over and started hugging me. She was tumor free, she told me. She was looking forward to the rest of her life. Two years later Sharon sent me a photo of her first child. It's amazing how often I now meet people like Sharon who were helped by this new drug therapy. I still tear up when I meet them. But that's what my work is ultimately all about, and why I am so committed to advancing immunotherapies."

Sharon has remained cancer free now for twelve years, "which by itself is phenomenal," says Wolchok. "However, the day that Sharon called to tell me that she was pregnant with her first child was the day that I finally told myself that I could now retire and feel good about my contribution to the world."

CHECKPOINT INHIBITORS BECOME GO-TO TOOLS IN THE FIGHT AGAINST CANCER

Ipilimumab was the first in a class of drugs known as checkpoint inhibitors to be approved against cancer, but it is certainly not the last. Jimmy Carter's pembrolizumab (KEYTRUDA) is another example. It works

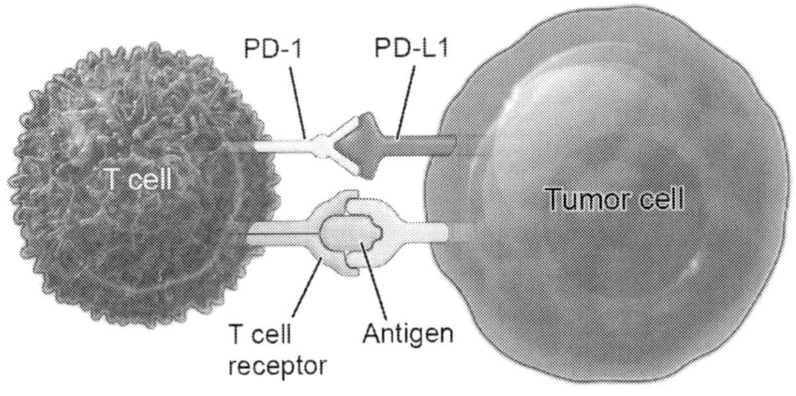

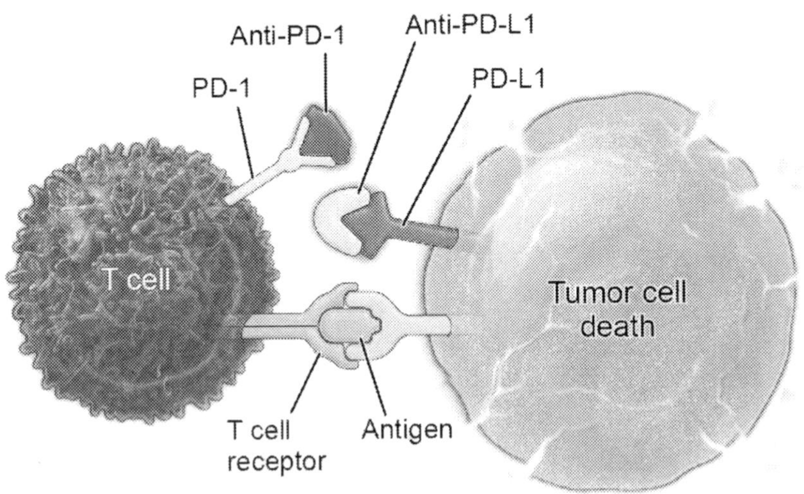

Another way cancer cells render themselves invisible to the immune system (top) is by waving a marker (PD-L1) on their surface that attaches to a complementary receptor on T cells, telling them that they are harmless.

Drugs that block the PD-1 receptors on either the T cell or the cancer cell (bottom) trip the cloak of invisibility from the tumor cell, allowing the T cell to kill it.

not by removing the brakes on the immune system, but by applying a blindfold. Again, the system depends on T cells. Think of a T cell as an alien in a science fiction movie, a creature with an eye at the end of an antenna. Only instead of using this eye to watch for things to zap with its ray gun, the alien watches for things to *not* zap. Technically, the "eye" is a protein called PD-1. It watches for its counterpart, PD-L1, on other cells. PD-L1 is like a flag that identifies a cell as a "friendly." The trouble is, many cancer cells also wave this PD-L1 flag. When a T cell's PD-1 sees a cancer cell's PD-L1, the T cell holds off its attack.

This interaction between PD-1 and PD-L1 is a "checkpoint," one of the ways the immune system knows not to attack the body it patrols. (When it does, that's autoimmune disease.) Pembrolizumab (KEYTRUDA) is a checkpoint inhibitor. By binding to PD-1, it covers T cells' "eyes," blinding them to cancer's friendly flag. This works because healthy cells use many checkpoints to deactivate immune system aggression, whereas some cancers (especially melanoma) are dependent on only this one PD-1 and PD-L1 checkpoint, so using pembrolizumab to blind T cells results in the death of tumor tissue usually with very little damage to healthy cells. There are new immune therapies on both sides of the PD-1/PD-L1 divide, some that blindfold T cells and others that strip away cancer's protective flag. Either way, when one fails to see the other, T cells attack.

THE EMERGENCE OF CAR-T THERAPY

Now we've seen two ways to release the immune system against cancer cells. But what happens when the immune system is naturally unable to recognize cancer cells? We already mentioned this as a major challenge in cancer immunotherapy. Because cancer arises from the body's own tissues, it can look to the immune system like a friend rather than an enemy. In this case, even with the brakes removed, the immune system wouldn't know where to attack. But there are chinks in cancer's disguise, minute differences that mark cancer cells as different from healthy ones. One of the earliest of these small but important differences was noticed in leukemia.

The word *leukemia* comes from the Ancient Greek words for *blood* and *disease* and applies to the entire range of blood cancers. Specific types of leukemia have to do with which type of blood cell has gone bad. In leukemia, dysfunctional blood cells from the bone marrow multiply prolifically, crowding out healthy blood cells. Unlike a solid tumor, leukemia is diffuse—there is no way to remove it surgically. When first-line treatments, usually chemotherapy, fail, doctors may resort to completely destroying the cells within a patient's bone marrow via radiation (as discussed in chapter four) or with a more aggressive chemotherapy and then replacing it with stem cells transplanted from a matched donor. When a matched donor can't be found or when that last-ditch treatment is followed by a relapse, patients are left with little hope for recovery.

That was the case in 2010 for Bill Ludwig, a retired corrections officer from Bridgeton, New Jersey. Bill, age sixty-five, had already battled his cancer for nine years; chemotherapy had held his chronic lymphocytic leukemia at bay. But the disease had finally pushed past the barricades of chemotherapy, and his options were scarce.

Like Sharon Belvin, Bill had nothing to lose. He enrolled in a phase I clinical trial at the University of Pennsylvania's Abramson Cancer Center. Though he couldn't be sure he would benefit from the trial, he hoped he could help doctors learn lessons that could benefit future patients. And like Sharon, Bill would be patient number one. This therapy had never been tested in humans.

In the trial, led by Carl June, MD, and David Porter, MD, doctors would harvest T cells from Bill's body. June and his team would then add a receptor to these T cells that would reprogram the cells to seek out the protein CD19, which is expressed in certain types of leukemia. These engineered T cells would then be infused back into Bill's body in hope that they would recognize and attack his leukemia. The strategy is called chimeric antigen receptor T cell, or

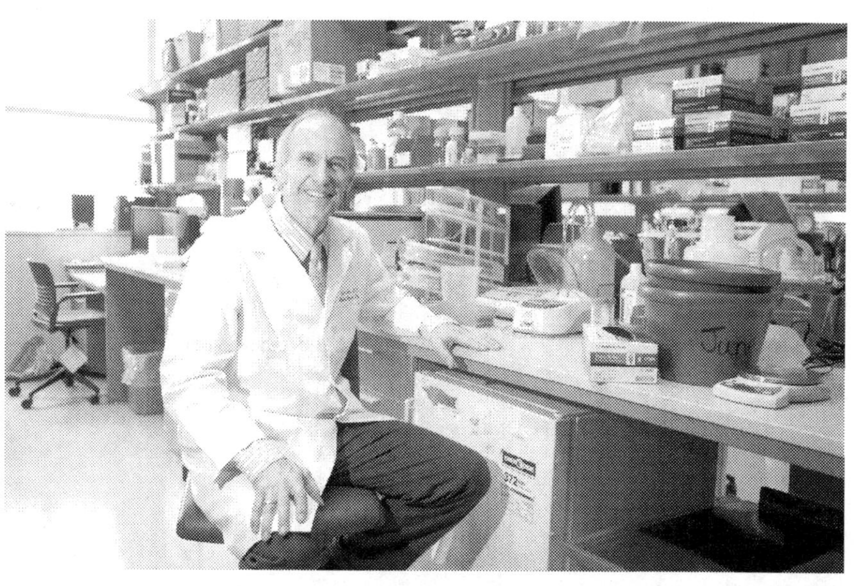

Cancer immunotherapy researcher Dr. Carl June. (Courtesy of Penn Medicine)

CAR-T, therapy. June had done the pioneering research behind this technique over more than a quarter century, beginning with his work as a physician-scientist studying HIV and AIDS in the US Navy, and later as the director of translational medicine at the Abramson Cancer Center at the University of Pennsylvania. "It's easy to cure mice with cancer drugs," June admits, "but when you go to humans, the experimental drugs often don't work. And cancer has been the worst in predicting later success in human trials. We didn't know what to expect with our CAR-T."

At Penn, the research team drew blood from Bill's arm and ran it through a sophisticated machine that extracted about a billion T cells. At a nearby lab, these cells were genetically engineered to express CD19 receptors and then grown in a special mixture that encouraged them to replicate and flourish. After this complex process, this "living drug" was introduced into Bill's body by a simple intravenous infusion.

Ten days later, Bill was hit with what felt like the worst flu of his life. His temperature skyrocketed, and he was racked with chills and

Opening of the Novartis-Penn Center for Advanced Cellular Therapies at Penn Medicine, from left to right: Glenn Dranoff, MD, PhD (Novartis); Bill Ludwig, Bruce Levine, PhD (Penn Medicine); and Carl June, MD (Penn Medicine). (Courtesy of Penn Medicine)

shakes. June and the rest of his doctors didn't know what had happened. After all, Bill was the first patient to try CAR-T therapy. Was he having an adverse reaction to these genetically engineered cells? Bill became so ill that he required care in the hospital ICU.

However, as doctors would learn over the course of the drug trial, Bill's symptoms were evidence of CAR-T cells doing their work almost too well. Inside Bill's body, these CAR-T cells continued to rapidly multiply as they attacked his leukemia. These engineered T cells were killing so many cancer cells that Bill's body struggled to handle the tremendous destruction of cells.

But then, Bill's condition improved, and where lymph nodes under Bill's arms had been almost bursting with cancer cells, the swelling had now diminished so that the nodes were difficult to feel. Thirty days after Bill was injected with CAR-T cells, when the results of Bill's pathology report came back, they didn't seem to make sense. June ordered another report. The second one confirmed the first: There was no evidence of leukemia in Bill Ludwig's body.

June estimated that each CAR-T cell had killed at least 1,000 tumor cells after being infused. The treatment had eradicated at least *two pounds* of cancer cells from Bill's blood.

For the first time in a decade, Bill was cancer free. In 2016, six years after receiving this pioneering treatment, Bill attended a ribbon-cutting ceremony for a new $27 million treatment center at Penn that specializes in delivering the CAR-T cell therapy that saved his life.

CANCER VACCINES IN THE CLINIC

In the five years since Bill Ludwig's treatment, several creative new strategies have been added to the field of cancer immunotherapy, including therapeutic vaccines made specifically to fight a person's individual cancer.

Some of these vaccines indirectly prevent cancer from developing, meaning that they ward off viruses that may give rise to cancer. These cancer-prevention vaccines include the hepatitis B vaccine, which staves off liver cancer, and the human papillomavirus vaccine GARDASIL 9, which guards against certain strains of HPV that lead to cervical and anogenital cancers.

However, unlike these preventive vaccines (and the vaccines against childhood diseases we're all familiar with), most cancer vaccines are not designed to prevent cancer in healthy people. Rather, these therapeutic vaccines are meant to awaken the immune system to the presence of a cancer, so that it will seek and destroy the cancer or at least slow its development. They work by introducing enough cancer-specific antigens into the body to push the immune system to manufacture more T cells equipped to target cells marked with these antigens. The first therapeutic cancer vaccine, PROVENGE, was approved for the treatment of prostate cancer in 2010, and international clinical trials for another prostate cancer vaccine, PROSTVAC, are now in the late stages of a phase III. Vaccine trials for breast cancer, melanoma, and brain cancer are also ongoing.

Still, other drugs use the immune system strategically to attack cancer cells directly. Medicines called monoclonal antibodies (MABs) use engineered antibodies to find cancer antigens, either acting against cancer cells alone or bringing with them payloads like chemotherapy molecules or radioactive particles that impede cancer growth by destroying cancer cells. There are several MABs already in use, including bevacizumab (Avastin), trastuzumab (Herceptin), and rituximab (RITUXAN), each of which is used to treat multiple kinds of cancer. With MABs, the challenge is finding the proteins that uniquely mark cancer cells. For example, some breast cancer cells coat themselves in the protein HER2; trastuzumab is designed to attach to HER2, silencing its ability to drive tumor growth. Another MAB, brentuximab recognizes the protein CD30, which marks many Hodgkin's lymphoma cells. In this case, the chemotherapy drug MMAE is attached to the MAB, which delivers the drug to the cancer cells. Clinical trials of many new MABs for additional cancer types are ongoing, and hundreds of new studies are recruiting patients. (To find these, search ClinicalTrials.gov for "monoclonal antibody" with the term "cancer.")

Now with an arsenal of techniques bolstered not merely by anecdote but by the very real positive data from widespread clinical trials, cancer immunotherapy is more commonly being considered as a mainstream first-line therapy for earlier stage cancers. Immuno-oncology is now used not only as a last-ditch effort for patients like Sharon Belvin and Bill Ludwig but also as the best treatment the world of medicine has to offer for patients like Jimmy Carter. In 2014 *Science* magazine named cancer immunotherapy the "Breakthrough of the Year," and at this writing, harnessing the immune system to fight cancer is the basis for over 3,400 human clinical trials in the United States and abroad. Dr. William Coley's nineteenth-century dream is now a reality. We are living in the age of immunotherapy.

CHAPTER SIX

STOPPING AUTOIMMUNE DISEASE

We carry our worst enemies within us.

—CHARLES SPURGEON

One night in 1991, Doug Melton's six-month-old son, Sam, started vomiting. For several days, Sam had seemed listless, but Doug knew that kids get stomach viruses, and he expected this one to pass like any other. It was November, after all. But now Sam couldn't stop throwing up, and soon it went from bad to worse. Sam went limp in Doug's arms. At that point, Doug and Gail, Sam's mother, rushed him to the emergency department at the Boston Children's Hospital. Doctors soon told them that Sam did not have a stomach flu and that his life was in danger.

This was one of the best pediatric hospitals in the country, and after running a series of tests, no one could figure out what was wrong.

"He was in very bad shape, and no one knew what to do," says Melton. "But just when Sam was about to take what we all thought was going to be his last breath, a nurse suddenly came in and told us that she had checked Sam's urine."

It turned out that Sam's urine was thick with the fatty acids called ketones. Instead of burning sugars, his body had switched to burning his limited fat stores for energy, releasing ketones as a by-product. As these ketones accumulated, his blood became more and more acidic until Sam reached extreme ketoacidosis. He was at risk of slipping into a coma and from there into death. All this was because poor Sam's immune system had mistakenly attacked and killed the cells in his pancreas that make insulin. Without enough insulin to allow his cells to absorb glucose, he couldn't burn sugars to power his metabolism. Sam had type 1 diabetes.

Doug and Gail were stunned, and so were the doctors. These pediatric medical experts had been taught that juvenile diabetes, as it was called then, developed when a child was about seven years old and that diagnoses peaked at about age fourteen. "They had not been looking for diabetes in a six-month-old," Melton says. In fact, Sam was the youngest baby ever diagnosed with diabetes in the 122-year history of Children's Hospital.

The medical team immediately went into action, whisking Sam off to the intensive care unit, inserting an IV into his tiny arm, and infusing him with insulin. A day later, Sam's recovery seemed miraculous. He was once again happy, healthy, and nursing normally. Thanks to the doctors' quick action, Sam had been snatched away from the precipice of death.

Doug Melton, PhD, decided to take action, too. "I was not going to stand for this. I had to figure something out for my child," says Melton, who is a tenured professor in the Harvard University Department of Stem Cell and Regenerative Biology. After Sam's diagnosis, Melton turned his entire academic focus to diabetes and started gathering lab assistants who would help him come up with a cure for his son.

UNDERSTANDING DIABETES

In type 1 diabetes, the problem is the lack of insulin. Specifically, the body's immune system mistakenly recognizes the insulin-producing beta cells in the pancreas as foreign and attacks them. As the number of beta cells decreases, so does the amount of insulin produced. Over time, as more and more cells are killed off, the body is left with little insulin and thus little ability to use glucose as fuel for cells. Instead, this sugar is left to circulate uselessly, and eventually harmfully, through the blood.

There is no cure for the disease, but the condition can be controlled by carefully monitoring blood sugar and injecting insulin, manufactured by genetically engineered bacteria, as needed. However, while effective, this treatment is time-consuming, uncomfortable, and costly. It also requires constant vigilance. The consequences of missing just one measurement or injection can be severe. And even the most conscientious type 1 diabetic can't control blood sugar with the extreme precision of a normally functioning pancreas. Long-term repercussions

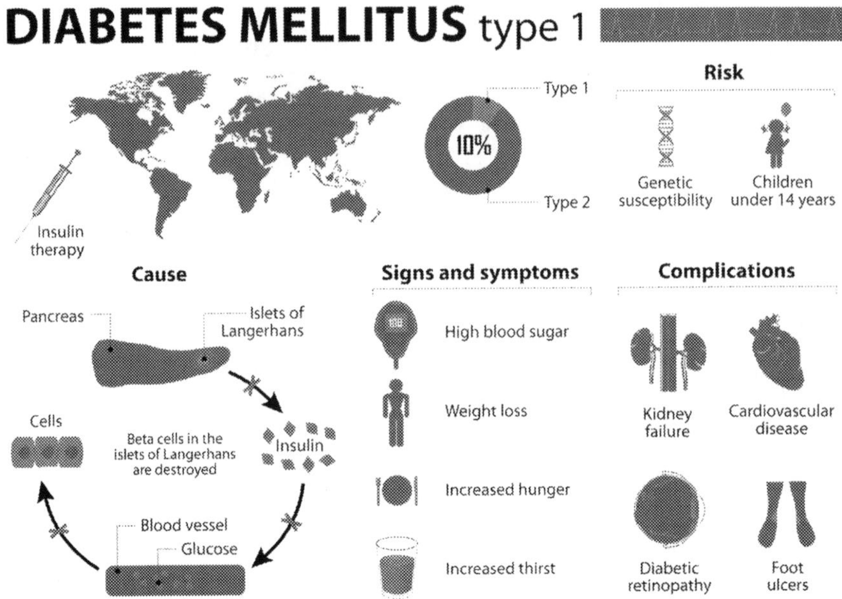

of mismanagement include kidney failure, circulation problems, heart disease, stroke, nerve damage, and amputations. Type 1 diabetes is also the leading cause of new cases of blindness among adults.

SEARCHING FOR A SOLUTION

Emma, Sam's sister, was also diagnosed with the disease when she was fourteen, and Dr. Doug Melton has been searching for a better way to treat patients like both of his children.

"My research goal has always been to replace insulin injections with nature's own solution through stem cells," says Melton. "A significantly better and more precise diabetes treatment would be to replace a person's destroyed pancreatic islet cells with healthy cells that could take control of the glucose monitoring and insulin release around the clock, without the person ever being aware of it."

Working with colleagues at Harvard, Melton's strategy is to use stem cells to grow new insulin-producing cells, and then to contain these cells in what is effectively a second, supportive pancreas. Progress is promising. In 2014 Melton reported in the prestigious journal *Cell* a thirty-day, six-step procedure for making hundreds of millions of functioning pancreatic beta cells in the laboratory.

"These beta cells are so finely tuned to their purpose in the body that I don't believe their function will ever be reproduced by people injecting insulin or by a pump injecting that insulin," Melton says. "In special diabetic mice, these novel beta cells cured their diabetes in less than ten days."

A few major challenges remain. Chief among them is how to keep the immune system from attacking and destroying these new cells as soon as they are placed in the body. A 2016 report in the journal *Nature Biotechnology* describes a creative strategy. Melton encases new insulin-producing cells in hydrogel capsules that act a bit like tiny shark cages. The semi-porous capsules allow glucose in so the cells can sense the variable levels of blood sugar, and they then let insulin produced by the cells inside the capsule escape into the blood. More important, the capsules keep immune attacker "sharks" safely outside.

Melton uses another analogy. "I liken our invention to a tea bag," he says. "The beta cells are in the bag and sense blood sugar, which goes into the bag. Insulin is then made and secreted out through the tea bag into the blood when needed. Nothing can get in that bag that could cause problems. The mice we tested this on were able to maintain proper blood glucose control for 174 days, which is decades in terms relative to the human life span."

Call it what you will, this tea-bag or shark-cage approach has the potential to provide people with type 1 diabetes a functioning artificial pancreas, freeing them from having to constantly monitor their blood sugar and take insulin and drugs. Melton is now gearing up for first-in-human clinical trials, in which a credit card–sized semi-permeable packet of stem cells will be implanted just under the skin in a simple surgical procedure.

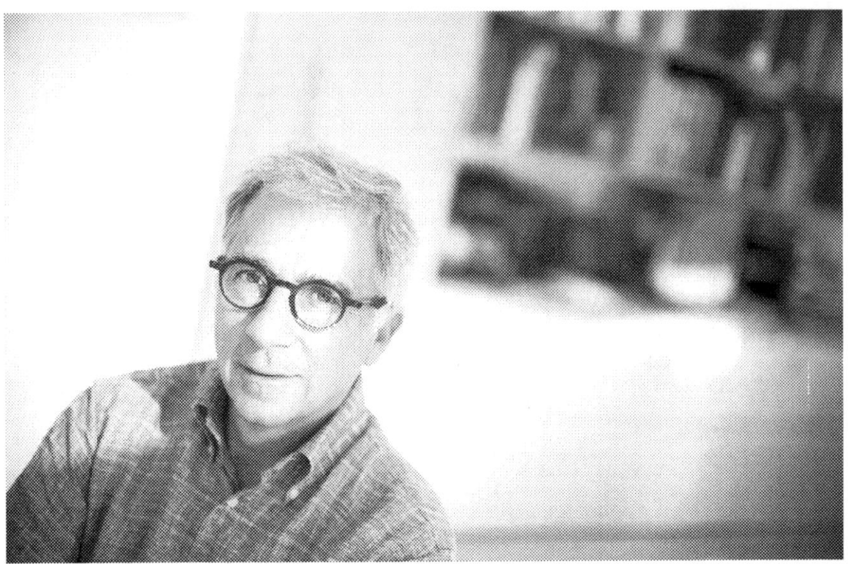

Type 1 diabetes researcher Dr. Douglas A. Melton. (Courtesy of Douglas A. Melton)

"I am looking forward to the day when my kids will finally be able to tell me, 'I used to have type 1 diabetes,'" Melton says.

OTHER AUTOIMMUNE CONDITIONS

Unfortunately, type 1 diabetes is only one of more than eighty recognized autoimmune diseases. The National Institutes of Health estimate that about 23.5 million Americans live with some form of autoimmune condition. The problem is the immune system's misinterpretation of tiny protein markers that coat the cells of your body as well as foreign cells. Because all your body's cells are built from identical genetic material, they should all express combinations of these cell-surface proteins that mark them as belonging to your body. In most cases, this self-labeling keeps the body's cells safe from attack by the immune system, which constantly patrols your body for foreign invaders. One of the most perplexing questions in medical research is why this system sometimes goes rogue. Why does the immune system misinterpret cell-surface proteins as foreign and attack the tissue that it is designed to protect?

For example, as you've seen, in type 1 diabetes, the immune system attacks insulin-producing cells in the pancreas. If you have rheumatoid arthritis, the immune system mistakenly attacks the lining of joints. In myasthenia gravis, autoimmune disease clogs important connections between nerves and muscles. And in multiple sclerosis, the immune system attacks the insulation surrounding nerve fibers in the brain and spinal cord.

MULTIPLE SCLEROSIS AND THE IMMUNE SYSTEM

"I was diagnosed with MS in the spring of 2007, when I was fourteen and a half," Roxane Beygi told an audience at the Vatican during the Second International Conference on the Progress of Regenerative Medicine in 2013. "At that time, I could hardly walk. I was told again and again that I should be in a wheelchair."

Roxane had many symptoms: dizziness, fatigue, and difficulties swallowing, speaking, writing, drinking from a glass, and more. "I'm not kidding—one of my major concerns at that time was not poking my eye out with a spoon or fork," she says.

Conventional treatments only added to her symptoms, bringing depression, headaches, and pimples. "I couldn't see how my conventional treatment had helped me. If anything it seemed to have added to my misery," she says. "At that time, I really had no hope. I really couldn't see a future for myself."

The list of symptoms makes it sound as if Roxane's body had been taken over by some terrible virus. In fact, it was her body itself that was to blame. Roxane's diagnosis of multiple sclerosis meant that her immune system had turned against the fatty layer of myelin that coats neurons. If you think of your brain cells and nerves as electric wires,

Multiple Sclerosis—Demyelination

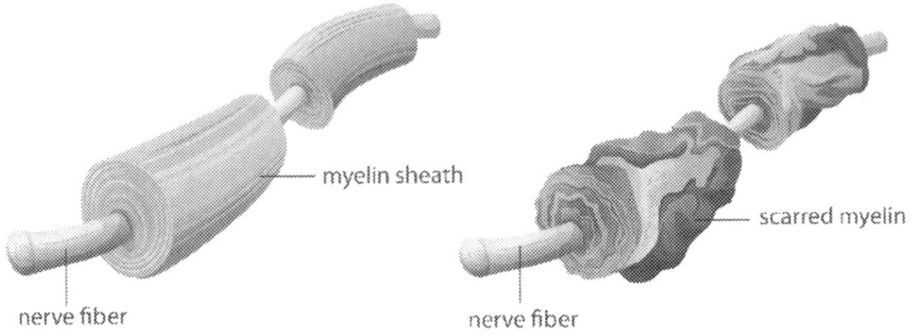

Multiple sclerosis (MS) is an autoimmune disease in which the body's immune system attacks its own central nervous system (the brain and spinal cord), damaging and destroying the myelin sheath, the substance that surrounds and insulates the nerves. The myelin destruction causes a distortion or interruption in nerve impulses traveling to and from the brain.

then myelin is the plastic insulation that keeps the electricity inside the wire. When myelin thins or gets patchy, electricity leaks from its proper passages, resulting in inefficient transmissions, short circuits, and, eventually, dropped signals. Resulting symptoms range from weakness to hallucinations to dramatic cognitive decline.

MS has two major stages. In the first, called relapsing-remitting MS, patients have flare-ups followed by periods of relatively normal function. It's as if something touches off an immune-system feeding frenzy against myelin and then the attack ends, often just as quickly. Neurons that survive these attacks may resume their original function, and the patient's symptoms may subside. However, neurons that are damaged past the tipping point and are killed do not regenerate. Even at this earlier stage, attacks can result in taking two steps backward followed by only one step forward, resulting in permanent symptoms.

Usually within ten to fifteen years after diagnosis, patients arrive at a stage called secondary progressive multiple sclerosis, in which "one step forward, two steps back" gives way to a slow, inexorable slide backward. Eventually, as myelin is eaten away, neurological decline and lack of muscle control become irreversible.

Richard Pryor is widely considered one of the most influential stand-up comedians of all time. He won five Grammy Awards, an Emmy, and the first Kennedy Center Mark Twain Prize for American Humor. This photo was taken several months before he was diagnosed with multiple sclerosis in 1986, and while he continued to perform for several years, he soon needed a wheelchair for mobility. Pryor died of a heart attack in 2005. (Photo by Alan Light)

About 400,000 people in the United States and about 2.5 million people around the world have MS. Approximately 200 new cases are diagnosed each week in the United States. While infections due to disability are common, suicide is the most frequent cause of death of MS patients.

As with most autoimmune disorders, while treatments may quiet the symptoms of the disease, there are no medicines that target the underlying cause of MS. Most current treatments seek to support myelin repair or reduce inflammation in the brain—a sign of the immune system's action. But what patients with MS really need is an immune system reboot, a way to force the immune system to start seeing myelin as friend instead of foe.

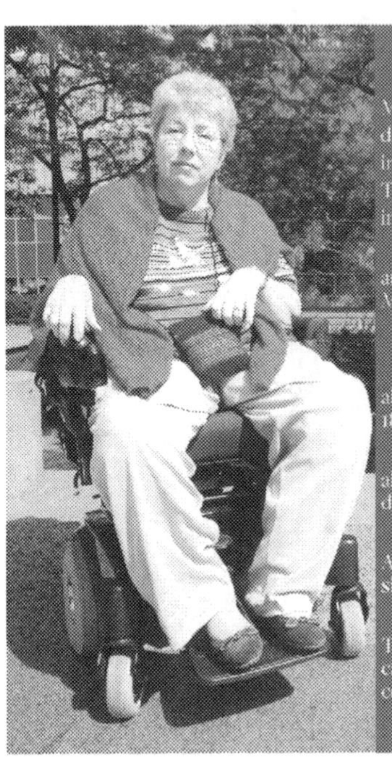

MULTIPLE SCLEROSIS

Multiple sclerosis (MS) is an unpredictable and potentially disabling disease of the central nervous system, which interrupts the flow of information within the brain and between the brain and body. The disease is thought to be triggered in a genetically susceptible individual by a combination of one or more environmental factors.

Anyone may develop MS, but there are some patterns. **Women** are at least **two to three times more likely than men** to develop MS.

About **250,000 to 350,000** people in the United States have MS.

Most people are **diagnosed between** the ages of **20 and 50**, although an estimated **8,000–10,000 children under the age of 18** also live with MS, and **people as old as 75 have developed it**.

Studies suggest that **genetic factors** may make certain individuals more susceptible than others, but there is no evidence that MS is directly inherited.

MS occurs in most ethnic groups, including African Americans, Asians, and Hispanics/Latinos, but it is **most common in Caucasians of northern European ancestry.**

An estimated **2.3 million people live with MS worldwide**. These numbers can only be estimated because MS disease activity can occur without a person being aware of it and symptoms may be completely invisible.

The National Multiple Sclerosis Society

DR. RICHARD BURT'S SEARCH FOR A STEM CELL MIRACLE

Richard Burt, MD, is working on an effective reset button for MS. His procedure is called an autologous nonmyeloablative hematopoietic stem cell transplantation. That mouthful of jargon means that he begins by harvesting immune stem cells from a patient's own blood. Then the patient undergoes low-dose chemotherapy to kill the majority of his or her white blood cells—the ones that are responsible for myelin attack. And finally, the previously extracted blood stem cells are reinfused into the patient, where they restore an immune system newly tolerant of the myelin it used to attack.

"These cells are easy to retrieve from the body," Burt says. "I can use hundreds of millions of the cells to make a new immune system for people with autoimmune diseases. The answer is stem cells, the primitive cells that give rise to all other cells in the body."

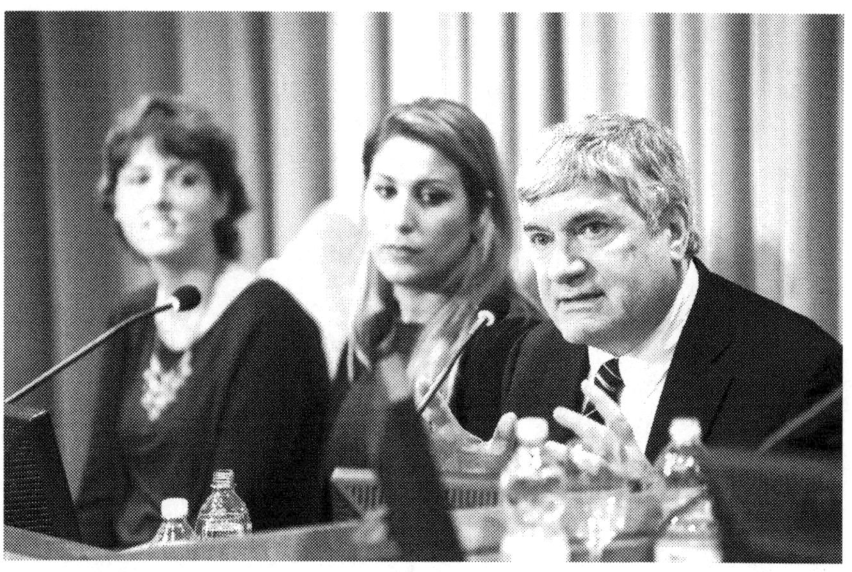

Dr. Richard Burt (right), Elizabeth Cougentakis (middle), and Grace Meihaus (left) at the Third International Conference on the Progress of Regenerative Medicine and Its Cultural Impact.

Between 2003 and 2014, 151 patients with relapsing-remitting MS were treated with this procedure at Northwestern University's Feinberg School of Medicine, where Burt is chief of medicine-immunotherapy and autoimmune diseases. Over the next few years after treatment, the volunteers were given multiple tests to measure their disability. The Expanded Disability Status Scale was used to measure cognition, coordination, and walking. Patients also underwent MRI scans and filled out extensive questionnaires to assess overall quality of life before and after treatment. In 2015 Burt reported the results in *JAMA*.

"In MS, the immune system is attacking your brain," Burt said. "After the procedure, it doesn't do that anymore."

The data agree. After chemotherapy and infusion of their own (i.e., autologous) blood stem cells, more than 80 percent of the patients in Burt's clinical trial did not relapse for the remainder of the trial (which seems much like a cure to people outside the medical field). More than half showed improvements in their disabilities, meaning that the procedure had not only slowed or stopped the progression of the disease but many patients were even able to recover function that had been lost.

Roxane Beygi was enrolled in this trial in September 2010. "I saw a miracle happen right before my eyes," says her mother, Evita. "Dr. Burt gave Roxane a second chance."

"Before, I had major fatigue where I couldn't even get out of bed," Roxane says. "Now, I get up at six to get ready for school. During the day, I'm busy with my studies, and I exercise regularly. Life is definitely much better. I have a lot of hope. I have a future now. Dr. Burt will always remain my hero."

"Current drug therapies for MS do not reverse disability or improve quality of life, and leave patients dependent on lifelong drug usage," says Burt. "However, if the results of stem cell transplants hold up in an ongoing randomized trial, it will fundamentally change the lives of patients suffering from MS."

REBOOTING THE IMMUNE SYSTEM WITH HEMATOPOIETIC STEM CELLS

Because Burt's technique essentially resets the immune system, it may be useful in treating autoimmune conditions far beyond multiple sclerosis. And Burt is indeed pushing the technique forward in a range of other conditions.

One condition for which autologous stem cell transplant is showing definite promise is systemic scleroderma.

When she was seventeen and a senior in high school, Grace Meihaus noticed that patches of her skin had suddenly become tight. Also, her fingers and toes swelled up and turned blue whenever it was cold outside. "That was really scary when it first happened," said Grace. "I didn't know what was going on with me. I felt different, and I didn't know why."

An active California teenager, Grace soon found herself tired all the time, struggling through exercise workouts that had once been

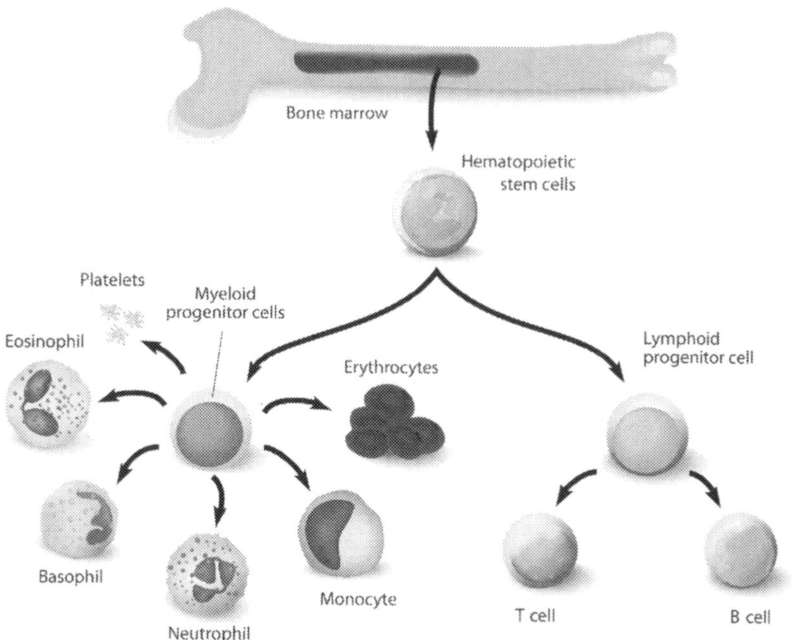

Hematopoietic stem cells can give rise to a variety of blood cells, including red blood cells and disease-fighting white blood cells of the immune system.

easy. A year later, her condition worsening, she made her way to a rheumatologist who quickly diagnosed her with systemic scleroderma. The ailment's name comes from the Ancient Greek *skleros,* meaning "hard," and *derma*, meaning "skin." Hardening of the skin is the most visible symptom, but when the condition is systemic, this hardening goes more than skin-deep, affecting other major organs of the body including the heart and lungs.

Grace was given a medication to ease her symptoms, but it didn't help. People with severe scleroderma often die within five years of diagnosis. "That really had me scared," she says. "I didn't know what to expect. My life now had a possible expiration date."

By 2015, as Grace's symptoms worsened, she became depressed and anxious, and her illness forced her to leave college. Through a scleroderma support group, she learned about Dr. Richard Burt at Northwestern and an experimental scleroderma study he was directing.

By the time Grace and her parents met with Burt, the disease had already started to harden her lungs, causing shortness of breath. "Knowing that I desperately wanted to go back to living a normal life, I agreed to join his study," Grace says. "I was extremely optimistic that everything was going to work out for me. My motto was 'hope dies last.'"

Several months later, Grace underwent Burt's stem cell "mini-transplant" procedure. After first harvesting stem cells from her blood, then administering chemotherapy for five days to wipe out most of her white blood cells and bone marrow, Burt then infused Grace's stem cells back into her body.

Within just a few days, Grace found that her skin had loosened, first on her hands and later on her face. "One day, I looked in the mirror and saw that I had laugh lines around my mouth," she said. "It was then that I knew that the stem cells were really working for me." Over the next few weeks, her energy returned, and her shortness of breath vanished. She could enjoy exercising again.

"I am very happy the way things have worked out for me," says Grace. "I feel normal, one year after the procedure. Five years after my first diagnosis, knowing that I am finally feeling better and the scleroderma is now under control, my life is back to normal, and I have a brighter future now. Even though the stem cell procedure was tough and difficult, it has all been worth it. It's remarkable when you think about it: my life was transformed by a little bag of my own stem cells, and I am grateful to be in good health once again."

MYASTHENIA GRAVIS: A LIFESAVING STEM CELL THERAPY

Grace Meihaus' stem cell story is amazing, but it isn't unique.

Elizabeth Cougentakis was healthy, athletic, and an outstanding student—until she turned thirteen. In that year, 2004, she was diagnosed

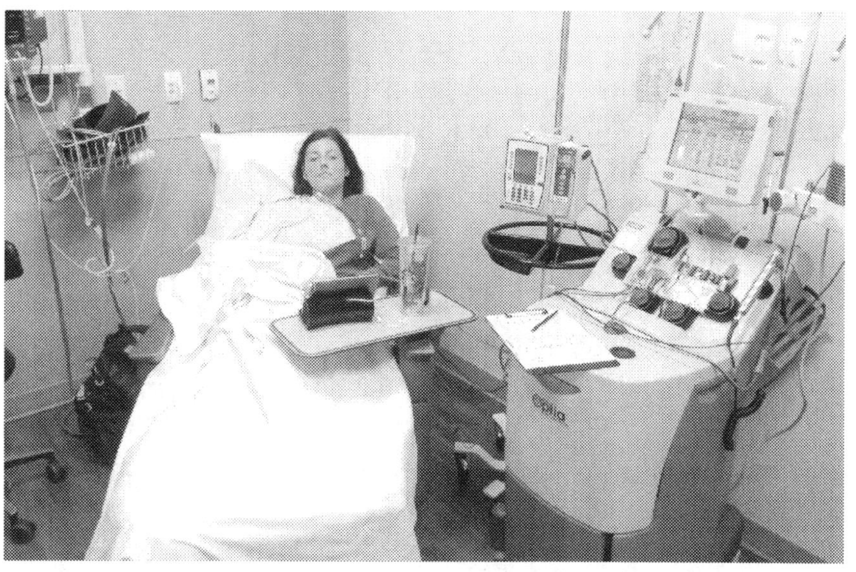

Grace Meihaus before treatment at Northwestern Memorial Hospital in Chicago, Illinois (July 2015).

Left: Grace during treatment with her father.
Right: Grace six months later, in January 2016.
(Photos courtesy of Grace Meihaus)

with myasthenia gravis, a serious neuromuscular autoimmune disorder that leads to muscle weakness and fatigue. In order to contract, human muscle cells need the stimulation of the neurotransmitter chemical acetylcholine. These cells have receptors on their surfaces that are a bit like the tentacles of sea anemones, waving in order to catch acetylcholine molecules. In myasthenia gravis, the immune system attacks and destroys these receptors, leaving muscle tissue unable to contract.

"The first thing I noticed was that when I smiled, my cheeks would droop and not go up. It looked like I was either upset or in pain," says Elizabeth. "Later, my eyes would droop and my vision started to blur. I started seeing double. Within a month, it had progressed to my arms and legs, and I lost all strength. I had trouble swallowing food . . . [and] I was choking on saliva and other liquids."

For most patients, daily medications can control the symptoms of myasthenia gravis. Elizabeth, however, was one of an unlucky few. Her condition continued to progress, and she quickly became completely disabled. Her breathing became difficult and labored, and two years after being diagnosed she needed a ventilator to breathe. Her parents began to feed her through a tube. Still, she continued to slip away.

After spending three months in an ICU, Elizabeth was finally sent home. Her doctors were perplexed. They told her parents they had never seen such a severe case of myasthenia gravis. They felt that she would be more comfortable with around-the-clock in-home professional care.

Over the next year or two, Elizabeth and her parents pursued several avenues, including having her thymus removed and traveling to Venezuela to explore experimental treatments, all to little effect. Then, in 2006 Elizabeth joined Dr. Burt's stem cell study at Northwestern. As in his treatment of other autoimmune diseases, Burt harvested stem cells from Elizabeth's blood, administered a short course of chemotherapy, then reinfused her stem cells.

Elizabeth had been on a feeding tube for two years, but after this treatment, her symptoms gradually lessened and finally disappeared. Within a year after Burt's procedure, she had recovered completely. For the last ten years, she has been completely healthy and requires no medication.

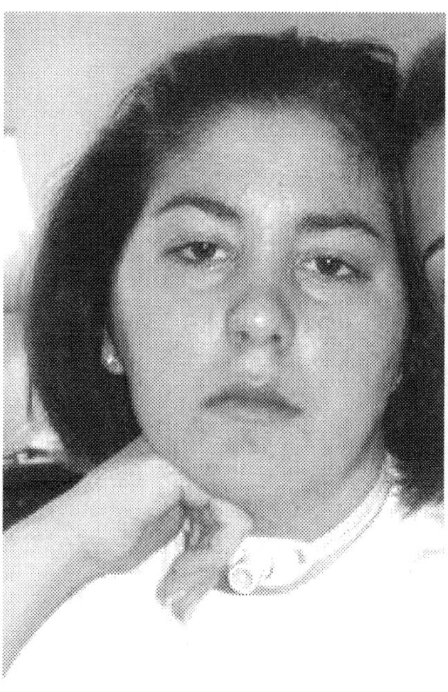

Elizabeth Cougentakis pictured before and after her treatment at Northwestern Memorial Hospital. Elizabeth was not able to control her facial muscles before treatment and, as seen in her later photo, regained facial expression after treatment. She has been healthy since her stem cell treatment in 2006. (Photos courtesy of Elizabeth Cougentakis)

Not everyone in Burt's trials makes a full recovery. Sometimes the immune system is able to reset itself, and sometimes it goes back to the same destructive behavior as before cellular treatment. This variability is one of the reasons the US Food and Drug Administration is progressing very deliberately toward approval of cell-based treatments for autoimmune conditions. But for Roxane Beygi, Grace Meihaus, and Elizabeth Cougentakis, autologous stem cell transplant is as close as medicine comes to a miracle cure. For more and more autoimmune diseases, in which the body's cells are the cause of a condition, they can also be the cure.

CHAPTER SEVEN

ALLERGIES AND FOOD SENSITIVITIES—PEANUTS, CELIAC, AND BEYOND

The best and most efficient pharmacy is within your own system.

—ROBERT C. PEALE, MD

In 2016 researchers from the University of Munich reported the case of a forty-six-year-old man who received a bone marrow transplant from his sister to treat leukemia. His body accepted the transplant, his disease went into remission, and everything seemed to be going well . . . until he ate a kiwi. After a couple of bites into the fruit, his mouth and throat started to swell until his breathing became ragged and he had to fight for air. Fortunately, the attack subsided.

MAJOR FOOD ALLERGENS

Each year, millions of Americans have allergic reactions to food. Although most food allergies cause relatively mild and minor symptoms, some food allergies can cause severe reactions and may even be life threatening.

While more than 160 foods can cause allergic reactions in people with food allergies, the Food Allergen Labeling and Consumer Protection Act of 2004 (FALCPA) identified the eight most common allergenic foods. These foods account for 90 percent of food allergic reactions and are the food sources from which many other ingredients are derived.

The eight most allergenic foods identified by the FALCPA are:

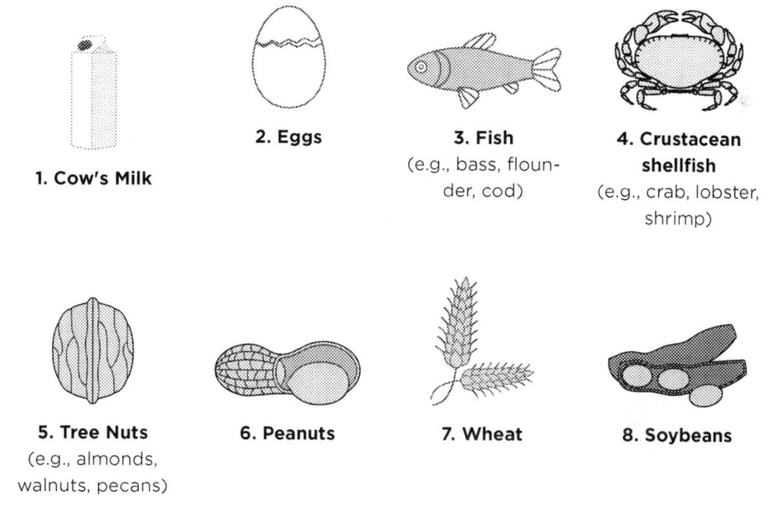

Food Allergen Labeling and Consumer Protection Act

It was inexplicable. He had eaten kiwis many times before and had never had an allergic attack, let alone something close to deadly. But his sister had. When researchers examined the patient's blood using a technique called fluorescence in situ hybridization, they discovered that it was his sister's donated blood cells that had initiated his immune system's allergic response. As described in the *Journal of the European Academy of Dermatology and Venereology*, the sister's bone marrow stem cells had saved the patient from leukemia. However, along with his sister's stem cells had come his sister's kiwi allergy, which almost killed him.

This, of course, shows how an allergy might be *caused*, rather than cured, by cells introduced as a medicine. But these are two sides of the same coin. If cell therapy could start an allergy, maybe it could stop one, too. An article in the journal *Blood* describes three such cases. In one, a patient received a bone marrow transplant from his sister to treat Fanconi anemia, a condition that affects the production of blood cells. Before the procedure, he had been allergic to latex. A reaction to latex during a surgery had once nearly taken his life. Twenty months after the bone marrow transplant from his sister, who was not allergic to latex, the man's level of blood cells specialized to attack latex had dropped to almost zero—he no longer had a latex allergy.

Another patient had suffered for twenty years from eczema, a bothersome skin condition. After he received a bone marrow transplant to treat leukemia, his eczema vanished. When yet another patient with eczema, asthma, and a peanut allergy was given a bone marrow transplant to treat a T and B cell disease, all three allergic

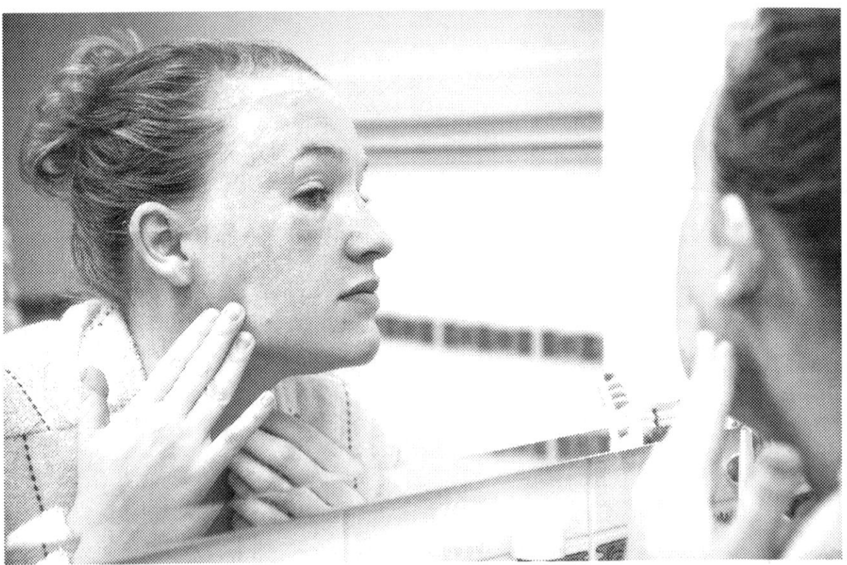

Eczema symptoms include itchy, red, and dry skin caused by inflammation, most commonly found on the face, hands, feet, inner elbows, and backs of knees. People with pollen, food, and other allergies are at increased risk for developing eczema.

conditions cleared up. The patient can now eat peanuts and peanut butter without any problems.

These cures for various autoimmune diseases were all the serendipitous side effects of bone marrow transplantation. To be clear, a bone marrow transplant would be a pretty extreme way to treat hay fever. It would take a tremendously debilitating case of allergies to justify the procedure. But the fact that swapping out the immune system cures allergic conditions suggests that other, less extreme approaches to immune-system manipulation might help protect people from dangerous conditions like asthma and the acute anaphylaxis of bee stings and severe nut allergies. A more targeted approach could even help people with itchy eyes and runny noses make it through pollen season a little more comfortably. Let's take a look now at some of these promising cellular therapies targeting allergies and asthma.

USING THE IMMUNE SYSTEM AGAINST ALLERGY

An allergy is a blunder of the immune system—an overreaction to something that should be nothing. That's why allergies can be caused or cured by transferring the immune system from one person to another via bone marrow transplant. Researchers are now following this trail deeper into the immune system to discover how we might manipulate it for the purpose of regulating allergic response.

If you've read the chapter on autoimmune disease, you already know a bit about the human body's system of defense. Basically, the immune system misinterprets the normal proteins that mark the body's cells as its own, seeing them as foreign or dangerous and mistakenly launching an attack. The same happens with allergies. The immune system sees proteins, in this case the antigens on peanuts or dust or pollen or any other environmental irritant, and launches an unnecessary attack. When the immune system overreacts in this way, you end up

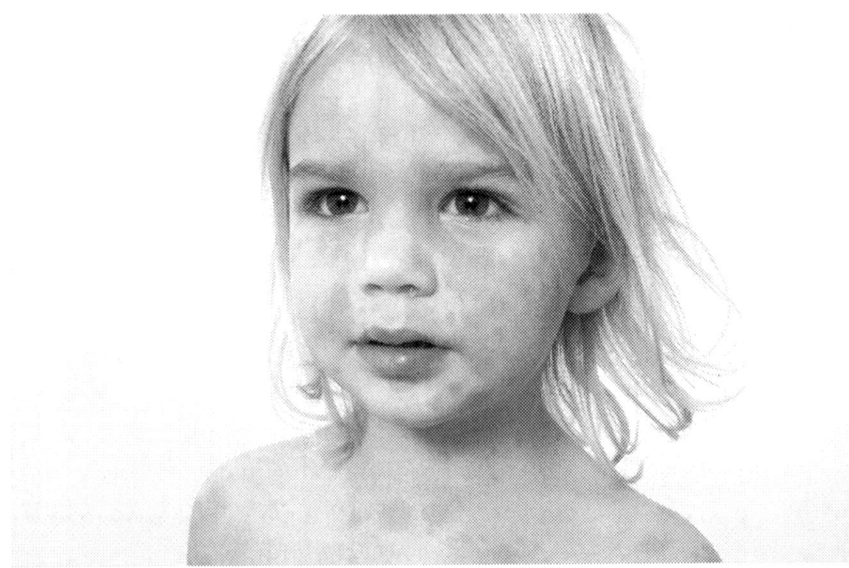

Hives are itchy, swollen, pale red bumps on the skin that suddenly appear as a result of an allergic reaction or sometimes for unknown reasons. The bumps can show up anywhere on the body and can last from a few hours to a day or more before fading.

with symptoms ranging from mild irritation to a life-threatening cascade of inflammation that swells your airways and shuts off your ability to breathe. With allergies, your immune system's erroneous attempts to protect you can kill you instead.

That's what happened to a one-year-old boy in California named Alex. Almost since birth, Alex had been colicky, and he had frequent bouts of eczema. When he got sick, he wheezed. This cluster of symptoms might have signaled allergies to a pediatrician, but it wasn't until Alex's first taste of scrambled eggs that his parents connected the dots. Within 30 seconds, he broke out in hives all over his body. His mother was horrified to discover they had no Benadryl or other antihistamine medications that could have stopped the allergic reaction. She called their pediatrician, who told her to watch Alex carefully for the next hour and to call 911 if his condition worsened. Luckily, the hives subsided, but the episode showed that Alex was in danger. Alex's parents had him tested for allergies. It turned out that he was not only allergic

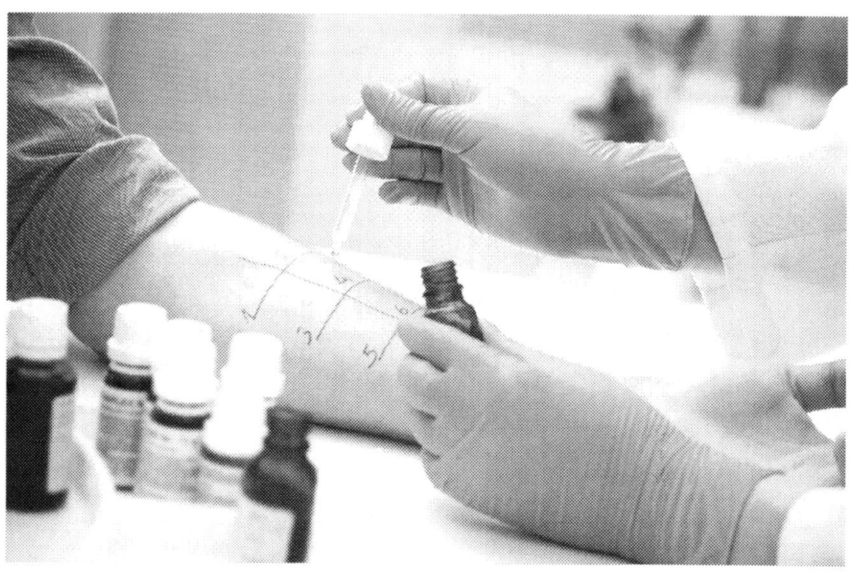

Skin testing for allergic reactions is usually done on the skin of the arm or sometimes the back. Several suspected allergens (such as pollen and pet dander) are placed into or onto the skin to look for red bumps that indicate an allergy to that compound. This type of test was inconclusive for Alex, which is why he needed a riskier food challenge test.

to eggs but also severely allergic to all types of nuts, a condition that would likely last his entire life.

"The feelings were overwhelming. How could we keep Alex safe?" his mom says.

Alex's parents did what any family would do—for years, they avoided foods that could cause a reaction. His mother spoke about how "daily life was scary and required tremendous energy and effort just to take care of Alex, educate family and friends, and avoid dangerous situations. We were limited in so many ways—reading labels *every* time and assuming they were accurate, no birthday cake, no deli foods, no buffets, no Halloween or Valentine's or Easter candy, limited restaurant options, no Chinese, Thai, or Indian food, and no ice cream cones, etc."

The family also kept their eyes on emerging research. It turned out that as Alex and his family were learning to live with his condition, doctors at nearby Stanford University were exploring a new strategy against allergies: Could a drug based on the inner workings of immune cells keep Alex's immune system from overreacting to peanuts or eggs or other allergens?

The strategy at Stanford depended on something called immunoglobulin, the antibody proteins that attack antigens. You've heard the saying "It takes a thief to catch a thief." That was the strategy. The researchers were testing a new kind of synthetic antibody, the drug omalizumab (XOLAIR), that latches onto a special kind of immunoglobulin involved in allergic asthma and pins its arms back, so to speak, preventing it from latching onto other antigens—peanut antigens, for example. Because this "trapped" immunoglobulin can't respond to antigens, it can't initiate a dangerous allergic reaction. At least that was the hypothesis.

Alex's mom remembers how terrifying it was to have her son, then six years old, subjected to the double-blind allergy test that was required in Stanford's clinical trial of the drug. Because skin and blood tests for allergy are notoriously inconclusive, Alex would be intentionally exposed to foods that could kill him. Neither Alex nor his parents nor the nurses doing the testing would know what was being tested that day. Of course, because the tests were in the controlled environment of a university clinical trial, doctors could quickly rescue Alex at the first sign of an allergic reaction. But still . . .

After the test, it turned out that Alex was indeed allergic to eggs, peanuts, cashews, and pistachios. As he started the trial, he was given tiny doses of the foods to which he was highly allergic, along with the drug modeled after and targeting the immunoglobulins that initiated his allergic response. The treatment worked for Alex, as it does for between a quarter and half of all patients who try the drug (as reported in *Annals of Internal Medicine* and the *Journal of Asthma* among many other sources). It reduced and even eliminated his symptoms, and, most important, almost completely removed the risk that an allergic immune response would go too far.

By August 2012, when he "graduated" from the study, eight-year-old Alex and his family had gone from afraid to empowered. "He has thoroughly enjoyed eating his 'firsts' so far: a peanut butter and jelly sandwich, French toast, brownies, dim sum, pancakes, dessert at a restaurant, and Grandma's famous pumpkin pie and Christmas cookies," his mom said. Alex continues on a maintenance dose of omalizumab (XOLAIR). He also eats eggs and nuts every day, which may also help maintain his desensitization. For Alex and an increasing number of people like him, cellular medicine was the cure for allergies.

ALLERGIES AND TREGS

Asthma is another kind of allergy, driven by the body's constant or acute overreaction to things like air pollution, pollen, or even cold air. And, as with allergies, researchers hope that cellular medicine used to regulate the immune system might hold a cure. One interesting strategy involves the use of regulatory T cells, or Tregs, which the body naturally uses to keep the immune system from overreacting to things that should be relatively harmless. Tregs provide balance—they're the cooler heads of the immune system. Immunoglobulin is a war hawk; it wants to help T cells launch a massive all-out attack against invaders, but Tregs are the voice of reason, suggesting that T cells reconsider before doing anything rash.

In fact, some types of asthma are defined by a reduction in the number or function of Treg cells. Doctors want to see thriving populations of Tregs producing proteins like CD25 and FOXP3 that tamp down the immune response, but many asthma patients lack enough Tregs for the task. For example, a study published in the journal *Pediatric Allergy and Immunology* reported that infants with food allergies have markedly low Treg levels, so the type of T cells that do the actual attacking (T effector cells) are left unchecked and overreact in ways that trigger asthmatic events. The fix seems simple: boost Treg function

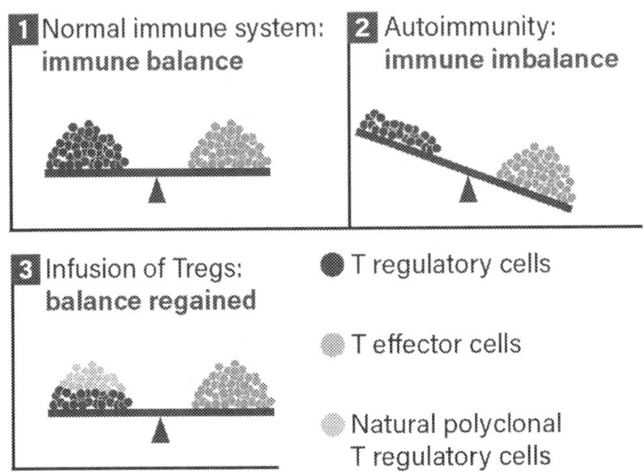

Immunomodulation: Autoimmune diseases arise and persist because of the loss of immune tolerance to one's own tissues. Immune tolerance is propagated through a variety of mechanisms, including suppression of self-reactive T effector cells and auto- or self-antigen-presenting cells by T regulatory cells (Tregs). The potential of Tregs as a therapeutic platform for different autoimmune diseases is in development. (Courtesy of Caladrius Biosciences, Inc.)

in people with asthma. But it turns out to be more complicated. There's no drug to fertilize the growth of Tregs and no easy way to amp up their activity.

Instead, one creative and somewhat counterintuitive technique is to intentionally provoke T cells. Seeing why this works requires us to take a couple of steps back. First, consider that allergy rates are on the rise, especially in first-world countries. Some researchers think a major reason for this is that we're simply too clean. Because of our constant washing, showering, and sanitizing, we're simply not exposed early in our lives to the range of allergens our ancestors were, so our immune systems don't have the opportunity to learn which antigens they shouldn't bother with. Researchers call this the "hygiene hypothesis:" an inverse relationship between the number of infectious agents in our environment and the incidence of asthma. The more we've sanitized our lives, the more people have developed asthma.

Because we no longer have as much opportunity for immune activity, our untrained immune systems sometimes act too much.

Again, the fix seems simple: just expose people to many antigens so that their bodies can learn to tolerate them—or else expose people to things that might stimulate enough Tregs to control immune overreaction. But this is challenging for obvious reasons. We can't just expose people to things that we know are harmful. Researchers have, however, exposed mice to many kinds of bacteria and parasites ranging from roundworms to *Toxoplasma gondii*, a protozoan that reproduces in cats and may affect the brains of other mammal hosts. In mice exposed to these antigens, Treg numbers and function go up. So what antigen could doctors give to asthma patients that would be strong enough to induce Treg formation but weak enough to do no harm on its own?

Inability to find a harmless antigen has led to creative, highly technical attempts to grow Tregs artificially. For example, tiny magnetic beads coated with special chemicals have been shown to make Treg populations blossom in the lab, and these cells could, in theory, be reinfused into the animal or person that gave the sample. But then the second problem emerges: How can you turn up Tregs without blunting the essential function of the immune system, which, after all, is to protect the body? By infusing an allergy sufferer with a massive dose of Tregs, will it make that person more prone to infections? The answer is something called antigen-specific therapy. Instead of promoting the growth of *all* Treg cells, doctors are learning to boost only the Tregs that tamp down asthma.

Right now, antigen-specific Treg therapy is at the stage of promising laboratory studies. For example, researchers grew Tregs specifically designed to block the immune system's attack on insulin-producing cells. They reported in the *Journal of Experimental Medicine* that these cells did a better job in mice of treating type 1 diabetes than regular, nonspecific Tregs. But this cellular approach to treating asthma and allergy is still far from the clinic. Fortunately, until there is a high-tech option, there exists a disconcertingly low-tech option.

ALLERGY AND THE MICROBIOME

The microbiome is the collection of microbes and microorganisms that inhabit your body, primarily the bacteria in your gut. Increasing evidence suggests that it's the composition of the gut microbiome that determines a person's risk of asthma and allergies. The American Academy of Microbiology estimates that you may have as many as ten times the number of nonhuman cells in your microbiome as you have human cells in your body. Let's look at this another way. Your DNA code includes about 23,000 genes. A study in the journal *Protein & Cell* shows that the genomes of the bacteria, viruses, and other microorganisms that make up your microbiome code for about 3.3 million genes. You like to think of yourself as human, but if you define your makeup by cell type, you are more precisely a walking petri dish containing over 100 trillion independent microorganisms, topped by a brain that believes it is in charge.

Fortunately, you live in harmony with the vast majority of the microorganisms that call you home. Actually, they are essential to our well-being. We live in a kind of constant balance with our gut bacteria. A healthy diversity of intestinal flora ensures that no single type of bacteria becomes powerful enough to do you harm. The messy balance of many microorganisms all working together, in parallel or often against one another, creates a kind of beneficial white noise that your body depends on for far-reaching aspects of your well-being. For example, the microbiota of the gut help you digest things that your body cannot digest on its own. The microbiota of your respiratory system and lungs protect against diseases like cystic fibrosis, asthma, and chronic obstructive pulmonary disease. The microbiota of the female reproductive tract protect against infection, and a mother's microbiota can affect her baby's immune system. Through these actions, your microbiome helps to regulate your metabolism and immune system, with trickle-down effects on nutrition, disease, and even the morphology of your body—that is, the way your body grows and changes shape.

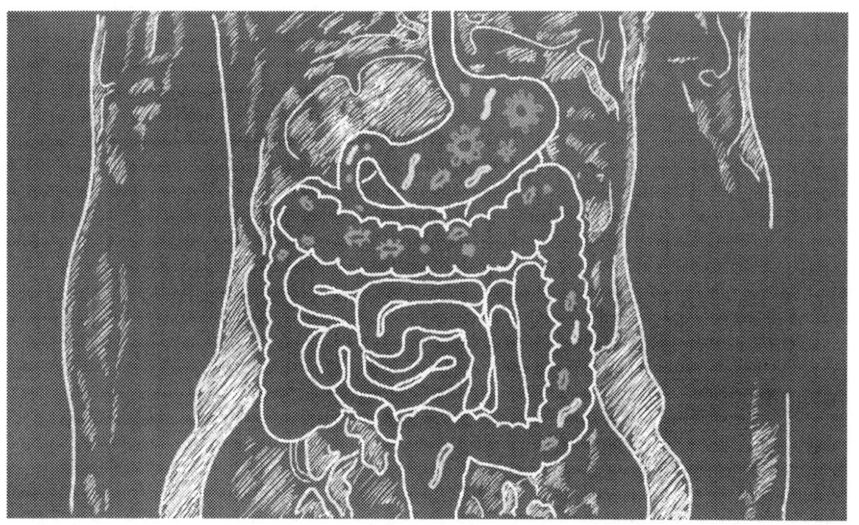

The human microbiome commonly refers to the bacteria that live in our intestinal tract. We are learning that these bacteria have a profound effect on the immune, endocrine, and digestive systems, and perhaps even obesity and mental health.

But when the microbiome becomes unbalanced, you have problems. The classic case is the bacterium called *Clostridium difficile (C. diff)*. Chances are you encounter *C. diff* frequently—many people carry the bacteria without symptoms, and many more overcome symptoms on their own without ever knowing what they had. That's because a healthy microbiome quickly outcompetes *C. diff*. But that can change if your microbiome becomes unbalanced. This sometimes happens to people who have recently taken large or repeated doses of antibiotics. In that case, your gut is a blank slate, ripe for colonization. And once *C. diff* gains a solid foothold, it can be difficult to uproot.

Another course of stronger antibiotics sometimes works. However, in some cases, small populations of *C. diff* that happen to be resistant to the antibiotic survive, and this hardy population can become entrenched. In that case, *C. diff* can become a chronic condition, leaving patients with symptoms ranging from mild diarrhea to life-threatening colon inflammation and bleeding resulting from *C. diff*'s attack on the walls of the gut.

The *New England Journal of Medicine* reports that in 2015 there were about half a million reported cases of *C. diff* infection and 29,000 deaths from the disease, usually as a complicating factor along with other conditions.

You have read here about cases in which blood disorders are cured by wiping out a patient's blood system and then replacing it with bone marrow stem cells from a matched donor. This therapy works in blood cancers and is gaining momentum as a treatment for autoimmune diseases. In the case of chronic, antibiotic-resistant *C. diff* infection, there is a similar, but somewhat yuckier treatment: replace the microbiome.

For *C. diff* that has resisted antibiotics, the treatment called fecal transfer is effective about 90 percent of the time. The treatment, unfortunately, is exactly what it sounds like. A donor's healthy stool, containing a healthy microbiome, is transferred to the patient's gut via a kind of enema in what can politely be called a poop transplant. This reset gives the healthy bacteria in the newly transferred stool the chance it needs to bring the *C. diff* under control and stabilize back into its natural equilibrium. Microbial transplant is now an FDA-approved therapy for all causes of colitis (inflammation of the colon), including, *C. diff*, Crohn's disease, ulcerative colitis, *Candida* (a genus of yeasts), and irritable bowel syndrome.

Surprisingly, there is increasing evidence that the influence of the microbiome goes beyond the gut. Some researchers have gone so far as to call the microbiome your "second brain," showing its influence on mood disorders, including depression and anxiety. This also seems related to the microbiome's effect on the immune system, which brings us back to asthma and allergy. For example, a study in *Nature Medicine* shows that an infant's gut flora at birth influence whether or not the child will go on to develop asthma. A healthy, diverse population of microbes in the gut ensures that the immune system can sort out what is and is not a big deal. On the other hand, a pattern of deficiency in four key species of bacteria makes a child *four times* more likely to develop asthma. The study also states, "Breastfeeding, vaginal births (as opposed to C-sections) and even having dogs in the household during the first year of life are all associated with protective effects against allergies and asthma." There is even some preliminary evidence that the composition of a person's gut microbiome is what predisposes them to obesity.

Research on the prevention side is clear: a healthy microbiome created in large part by interacting with an environment that isn't perfectly sterile goes a long way toward protecting kids from asthma and allergies—the "hygiene hypothesis" previously mentioned. But what about people who already *have* these conditions? It's an open question, but it's looking more and more as if the answer will depend on cells. For example, University of Chicago biologist Cathryn Nagler, PhD, showed that mice given antibiotics early in life were more likely to later develop peanut allergies. When she introduced certain species of the genus *Clostridium* to their guts, peanut allergies vanished. These results, published in the *Proceedings of the National Academy of Sciences,* pointed out that it wasn't all bacteria but members of this specific genus that eliminated the rodents' peanut allergies.

Nagler's study hints at how we might use cell-based medicines to target allergies. For example, all it takes is a quick trip to your local health food store to see that probiotics have captured the popular consciousness. Promises of the products include better sleep, a stronger immune system, more energy, and almost any other benefit you can imagine. But studies of probiotics against allergies, of which there have been many, have been inconclusive. So far, taking a general probiotic supplement hasn't been conclusively shown to reverse allergies or asthma. The key, as Nagler's research shows, may be our ability to match specific bacteria with specific conditions. Again, rather than a molecule that turns your body's machinery up or down, the cure for allergies and asthma may depend on finding the right cell to do this work for us. The same may be true in one of today's most visible diagnoses, gluten intolerance due to celiac disease.

GLUTEN INTOLERANCE AND CELIAC DISEASE

Celiac disease is a serious genetic autoimmune disorder in which the ingestion of gluten—a protein typically found in wheat, rye, and barley—leads to an immune response that attacks the small intestine. Eventually, the villi—small, fingerlike projections that line the small intestine and help promote nutrient absorption—can be damaged. Chances are you know someone who carries the genes for celiac disease. That's because 30 to 40 percent of the population carry small variations in genes that code for the HLA-DQ proteins. These slightly defective proteins create an immune response against gluten, spurring the immune system to attack the small intestine and leading to symptoms like diarrhea and fatigue. Yes, gluten intolerance is an autoimmune condition.

But it seems that while a third or more of all people carry celiac genes, only about 1 percent actually develop celiac symptoms. In addition to genetic predisposition, there must be something else needed to create the actual symptoms of the disease. Could it be the microbiome?

A paper in the *Annals of Nutrition and Metabolism* reported that people with celiac disease tend to have microbiome imbalance and people with microbiome imbalance *and* those defective HLA-DQ genes are more likely to develop symptoms of celiac disease. But, as we've seen, fixing a general imbalance might not be enough to reverse a mistaken immune response. Curing celiac symptoms may depend on finding a specific microbial culprit, and researchers from the Farncombe Family Digestive Health Research Institute at McMaster University in Hamilton, Ontario, have a good candidate. One of the bacteria that overpopulate the guts of celiac sufferers is *Pseudomonas aeruginosa*. At McMaster, researchers showed that mice with this bacterium in their guts metabolized gluten differently than healthy mice.

"So the type of bacteria that we have in our gut contributes to the digestion of gluten, and the way this digestion is performed could increase or decrease the chances of developing celiac disease in a person with genetic risk," said Elena Verdú, MD, PhD, senior author of the study and an associate professor of medicine for the Michael G. DeGroote School of Medicine at McMaster. "Celiac disease is caused by gluten in genetically predisposed people, but bacteria in our gut could tip the balance in some people between developing the disease or staying healthy."

If *P. aeruginosa* is the villain in celiac, then a paper in *Clinical Microbiology Reviews* spotlights the heroes. Comparing children with celiac to control groups showed that affected kids specifically lack lactobacilli and bifidobacteria. In a venue far from the research lab, lactobacilli help to break down and metabolize gluten to the point that some bread

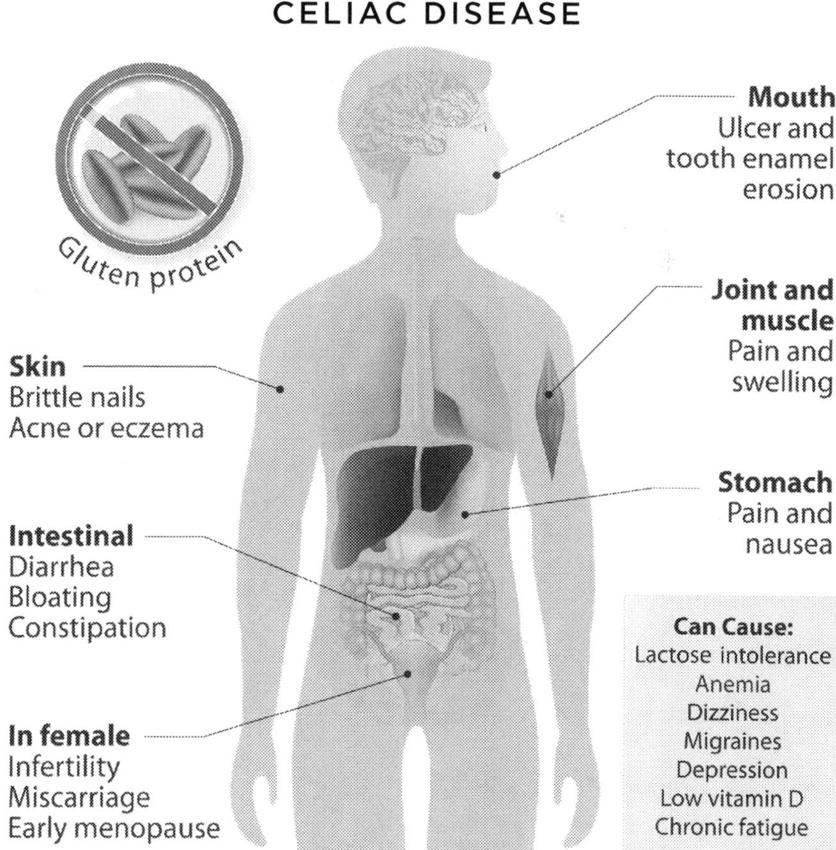

makers are experimenting with adding these bacteria to their dough to make a normally glutinous product gluten free.

For now, it is impossible to change the genes that predispose people to celiac disease (although in chapter nine, you'll learn about the hope offered by the gene-editing technology CRISPR). But could the cellular therapy of adding in these missing microbes help people with celiac genes avoid symptoms? For decades, medical science has been focused on eliminating the cellular and microbial causes of disease. Now we are discovering that some cells can be helpful. Our bodies did not evolve in a sterile environment, and while sanitation has certainly helped to protect us against many life-threatening infections, it also means that we increasingly fail to pick up important microbial collaborators from the unnaturally clean world around us. Recently, doctors and scientists have made progress toward reintroducing these essential bacteria cells to our bodies. The result is leading to new, cell-based treatments against allergy and asthma.

GLUTEN-CONTAINING FOODS

There are many food items that may contain gluten, so always read the label of any food product you buy if "gluten free" is not specified on the label.

Breads and Pastries
flatbreads, cornbreads, potato bread, muffins, donuts, rolls, pita, naan, bagels, croissants

Cereal and Granola
corn flakes and rice puffs that contain malt extract/flavoring, granola (while it is usually made with oats, which are gluten free, oats can be contaminated with wheat and other gluten-containing grains during processing)

Noodles
ramen, udon, soba (when made with only a percentage of buckwheat flour), chow mein, and egg noodles (Note: rice noodles and mung bean noodles are gluten free)

Pastas
raviolis, dumplings, couscous, gnocchi

Breakfast Foods
pancakes, waffles, French toast, crepes, biscuits

Crackers
pretzels, goldfish, graham crackers

Baked Goods
cakes, cookies, pie crusts, brownies

Breading and Coating Mixes
panko breadcrumbs

Croutons
stuffings, dressings

Sauces and Gravies
traditional soy sauce, cream sauces made with a roux, sauces using wheat flour as thickener

Beer, Malt Beverages, and Brewer's Yeast
anything using "wheat flour" as an ingredient, unless explicitly gluten free

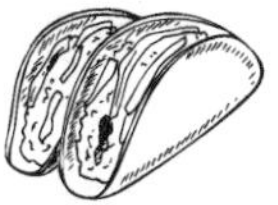

Flour Tortillas

Distilled Beverages and Vinegars
most distilled alcoholic beverages and vinegars are gluten free; these products do not contain any harmful gluten peptides; even if they are made from gluten-containing grains, research indicates that the gluten peptide is too large to carry over in the distillation process, leaving the resulting liquid gluten free;
wines and hard liquor/distilled beverages are gluten-free;
beers, ales, lagers, malt beverages, and malt vinegars that are made from gluten-containing grains are not distilled and contain gluten

Celiac Disease Foundation

CHAPTER EIGHT

THE FUTURE OF IMMUNOTHERAPY

Healing is a matter of time, but it is sometimes also a matter of opportunity.

—HIPPOCRATES

Where we are now covers about 2 percent of what immunotherapy will do for cancer," says Drew Pardoll, MD, PhD, director of the Bloomberg-Kimmel Institute for Cancer Immunotherapy at Johns Hopkins University. "We really are just scratching the surface."

The first step in achieving the other 98 percent, not to mention extending immunotherapy to conditions outside of cancer, is to use existing drugs better. It has taken more than a half century of experimentation to come to our current uses of chemotherapy and radiation, and we are still learning new ways to combine and sequence these older treatments to help patients live longer, better lives. The field of immunotherapy is just at the beginning of a similar path. We know that a handful of individual immunotherapies are effective, and the question is how to squeeze every last bit of benefit from their potential.

One way is to combine immunotherapies with existing chemotherapies. Stephen Estrada knows all about it. On the last day of a family trip to Jamaica, Stephen felt too sick to do anything but lie in his hotel bed. On the flight home to Colorado, he was in too much pain to sit and had to pace the aisle. After a trip to the emergency room, Stephen headed home with a diagnosis of gas pains or other gastrointestinal illness, probably from travel. But the pain remained. Eventually, he visited his primary care doctor who found that lymph nodes in his back were the size of peaches and ordered imaging.

"I was rushed into an emergency colon resection four days after my first CT scan showed a cancer-like mass," he says. "I was diagnosed with stage four colon cancer at the age of twenty-eight. Cancer changed everything about my life immediately. I felt like an altered version of myself."

After Stephen's disease continued to worsen on traditional therapies, his doctors at the University of Colorado Cancer Center helped him enroll in a phase I clinical trial of the immunotherapy atezolizumab (TECENTRIQ), which would be combined with the chemotherapy drug bevacizumab (Avastin). This proved to be a powerful one-two punch, with Avastin bringing his cancer to the brink and atezolizumab recruiting his immune system to push the cancer cells over the edge. The immunotherapy used in the Colorado trial is a PD-L1 blocker. As you read in chapter five, this checkpoint inhibitor stands in the middle of a handshake between a cancer cell and the immune system so that the two cannot form a truce. Without this truce, the immune system remains activated against cancer cells, killing them as if they were bacteria, influenza, or any other common invader. The five-year survival rate for stage 4 colon cancer is not good, only 11 percent, but this cutting-edge combination therapy helped Stephen beat the odds.

"My current status is that I have a stable disease," he says, after more than three years on the trial. "As for the future, I just want to live life fully. I want to be around the people I love. I want to be more honest with myself and others. I want to enjoy what I do for a living and just be thankful that I'm able to do anything. But most of all, I want the gift of immunotherapy to be accessible to everyone."

Stephen's wish is becoming a reality. In the future, immunotherapy will be combined with even greater numbers of traditional therapies, adding the power of the immune system to treatments that have worked well but not well enough in the past. Different immunotherapies will also be combined with each other. Mary Elizabeth Williams, a writer and mother of two living in New York, was one of the first to try out the idea that if one immunotherapy is good, then two might be even better.

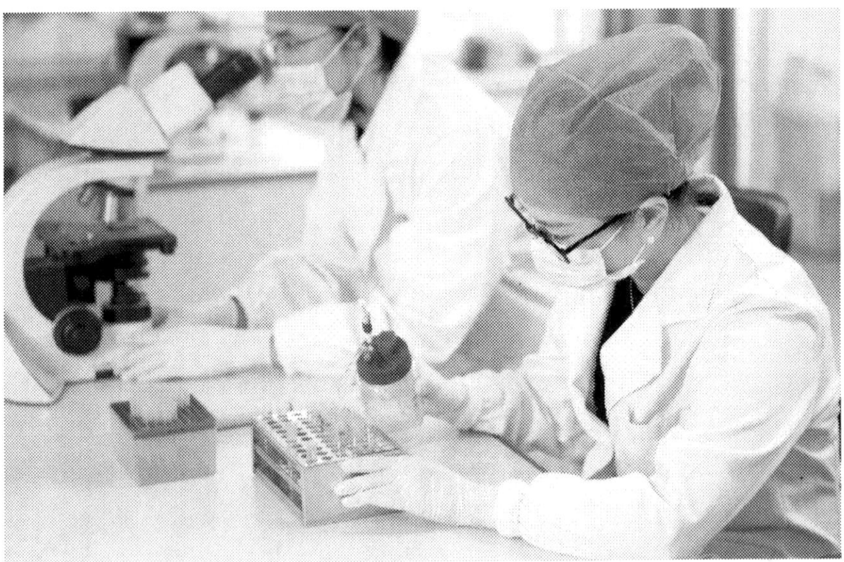

There are over 2,000 immunotherapy clinical trials underway and many more preclinical studies where researchers are searching for cures for cancer by using the body's immune system to destroy cancer cells.

COMBINED THERAPIES

"In the summer of 2010, I discovered that I had a little scab on the top of my head and I didn't think much of it for a while. I just assumed it was a scrape or a reaction to my shampoo or something. I went to my dermatologist and she said, 'That looks like skin cancer,'" Williams says.

Not only was it skin cancer, it was melanoma, the most dangerous form of the disease, and it had already spread throughout her body, resulting in a large tumor on her back and cancer cells in most of her organs. Williams was sure it was a terminal diagnosis.

"When my doctor told me that it was stage 4, my options dwindled down to a very, very precious few. We were going to try a traditional drug therapy and hope for the best, and then, fortunately not too long after my diagnosis, when I was flipping out about what I was going to do with the rest of my life, whatever that might entail, she called me and said, 'There is a spot in a clinical trial and I think you should explore this,'" Williams says.

The trial, conducted by Memorial Sloan Kettering Cancer Center, involved a combination of the anti-PD-1 immunotherapy nivolumab (OPDIVO) with another drug called ipilimumab (YERVOY), which is similarly designed to block a cancer cell's ability to hide itself with the protein CTLA-4.

"I signed off on a long list of possible side effects and held my breath to see what was going to happen. Luckily not a whole lot did. Negative side effects occurred. But what did happen was really exciting. I went back to the hospital one week later and my doctor looked at my tumor and it had started getting smaller. A few weeks later, I was out with a friend in a bar and he said, 'Can I see your tumor?' and I lifted up the back of my shirt to try to show it to him and he couldn't find it. It was really exciting," Williams says.

Today, Williams remains on a maintenance dose of these two immunotherapies so that her immune system stays vigilant against any possible recurrence. "What I'm looking forward to is the day that I'm playing with my grandchildren and they ask what it was like back when people got cancer and there wasn't much you could do about it," she says. "And I'm going to tell them that I was part of something that helped change that. That I was there at the beginning of the end of this."

The May 2015 issue of the *New England Journal of Medicine* reported the results of Williams' clinical trial. In all, 22 percent of patients with advanced melanoma saw what doctors call a "complete response," meaning that their cancer became undetectable. Let's rephrase that to make sure it sinks in: these patients were almost certainly going to die, and soon; they now have no signs of cancer.

It turned out that one secret driving this combination's success was buried deep in the genome of Williams' melanoma. It had what is called a BRAF mutation, which allows some cancer cells to grow and divide more quickly, but also makes them especially sensitive to immune system attack. Melanoma is not the only cancer harboring BRAF mutations, and this brings us to a third way that we will learn to use immunotherapies in the future. We will extend their use past the diseases they were designed to treat. For example, some lung cancers and some pediatric glioblastomas have BRAF mutations. Might these same immunotherapies work in conditions harboring similar genetic changes?

The answer comes emphatically from Donna Fernandez, who, after being diagnosed with stage 4 lung cancer in 2012, joined a clinical trial of nivolumab combined with chemotherapy. Remember, nivolumab was developed as a melanoma drug. What use could it be in lung cancer?

"When I entered the clinical trial, I really didn't have much hope for it to help me personally. I was really doing it because I thought my time on earth was probably pretty limited, and that I might be able to help future generations," she says.

But like Mary Elizabeth Williams' melanoma, Donna Fernandez's lung cancer was driven by a BRAF mutation. When nivolumab took the brakes off her immune system, her immune cells aggressively attacked the cancer cells and the tumors that had spread throughout her body all started to shrink. Her experience tells a story that has come to define modern oncology. Doctors are learning to define cancers less by where they live in the body and more by their genetic fingerprints. In the future, as this approach becomes more common, immunotherapies along with genetically targeted therapies will shatter the paradigm in which cancers are defined by the site in which they are discovered. Patients will have new hope based on the understanding that certain cancers are genetically similar, even though they happen to grow in different organs or tissues. So far, we have looked only at the future of the most widely used class of immunotherapies, the checkpoint inhibitors that target cancer's use of proteins like PD-L1 and CTLA-4 to hide from the immune system. But close behind checkpoint inhibitors are a couple of more up-and-coming classes of immunotherapy. For example, bispecific T-cell engagers (BiTEs) are a new class of immunotherapeutic molecules that enhance a patient's immune response to tumors by physically connecting T cells to tumor cells. One end of a BiTE molecule is designed to grab a T cell, while the other end is designed to grab a protein marker on a tumor cell. Bringing the T cell close to the malignant cell allows the T cell to kill the cancer cell.

Another pioneering approach is a new kind of cancer vaccine that works somewhat differently from the way we usually think of vaccines. According to the American Cancer Society, "Instead of preventing disease, [these vaccines] are meant to teach the immune system to attack a disease that already exists." The first cancer vaccine to earn approval was sipuleucel-T (PROVENGE). In this treatment, a patient's immune cells are removed and then exposed to a protein called prostatic acid phosphatase (PAP). This "teaches" the cells to target PAP, so that when they are reinfused into the patient they will seek out the PAP that marks prostate cancer cells.

A twist on this vaccine approach called CAR-T therapy is already saving lives in clinical trials.

CAR-T IMMUNOTHERAPY

"On Friday, May 28, 2010, just a few weeks after my fifth birthday, I was diagnosed with standard risk pre-b acute lymphoblastic leukemia (ALL) at Hershey Medical Center," writes Emily Whitehead on her website. At that point, Emily's chance of beating the disease with chemotherapy was strong, about 85 percent. Unfortunately, though chemotherapy beat back Emily's disease, she soon relapsed, and almost two years after her initial diagnosis, Emily's parents, Kari and Tom, decided to transfer her care from their local hospital to the Children's Hospital of Philadelphia (CHOP).

"Emily's parents came to see me at CHOP for a second opinion because they were concerned about the next step for their daughter," says Susan R. Rheingold, MD, medical director of the oncology outpatient clinic and an attending physician at the hospital. "After examining Emily, I told them that although she was very sick, Emily still might benefit from another round of chemotherapy, as her local oncologist had recommended."

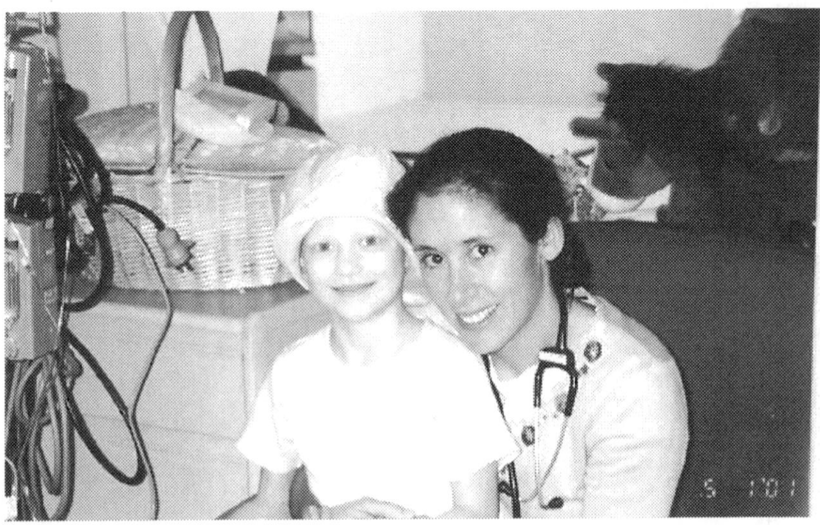

Dr. Susan R. Rheingold from Children's Hospital of Philadelphia is a leading researcher in pediatric leukemia. (Courtesy of Susan Rheingold)

Four months later, the news was frightening. Emily had relapsed again, and the Whiteheads reached out immediately to Rheingold. Twenty-four hours later, the family made the 215-mile trek to Philadelphia from their hometown of Philipsburg, Pennsylvania, and Rheingold outlined a plan that might help their very sick daughter. The idea behind the treatment, called CAR-T therapy, was to collect Emily's own immune cells, modify them so that they would be able to spot the cancer throughout her body, and then reinfuse these engineered immune cells into Emily's body to fight the disease. The study was part of a collaboration with researchers including Carl June, MD, of the University of Pennsylvania, who pioneered the use of CAR-Ts—chimeric antigen receptor T cells—in the treatment of B-cell leukemia. Several of June's adult patients had already been enrolled in the study and had responded well.

After doctors harvested Emily's T cells, they reengineered them in a laboratory using a disabled form of the HIV virus to insert a new gene in these cells' DNA. Her "new and improved" T cells would now make a protein called CD19 that is found only on the surface of B cells, the very cells that had become cancerous in Emily. Once the reengineered cells were infused back into Emily's blood, they would then be able to recognize the CD19 markers on the leukemic B cells that threatened Emily's life.

On April 17, 2012, Emily became the first pediatric patient in the world to be treated with the CAR-T immunotherapy then known as CTL019. A sign that this cellular therapy is working is what oncologists call "shake and bake." The patient becomes very ill, with uncontrollable chills and raging fevers, a sign that the therapy had indeed initiated a kind of civil war between the engineered T cells and the leukemic B cells. But as is sometimes the case with experimental treatments, Emily's symptoms were much more severe than doctors anticipated. She became critically ill and was admitted to the pediatric intensive care unit at CHOP and later placed on a ventilator to assist her breathing. At one point, she was very close to death and given only a one-in-a-thousand chance of survival.

Doctors worked tirelessly to determine what had caused Emily's sudden illness and soon learned that a certain protein—interleukin-6 (IL-6)—had become highly elevated, a sign of inflammation caused by her immune system's explosive attack. It was as if Emily's entire body had become inflamed. Interestingly, this same protein is involved in rheumatoid arthritis, another inflammatory condition. To treat rheumatoid arthritis, some patients take the drug tocilizumab (ACTEMRA) to block the inflammatory action of IL-6. Serendipitously—as sometimes happens in medical research—June's daughter has rheumatoid arthritis and uses tocilizumab as part of her treatment regimen.

Although the drug had never been used in this setting before, the team reasoned that it might dampen the IL-6 storm raging in Emily's body and immediately administered tocilizumab to Emily, with dramatic results. Her condition improved faster than anyone could have hoped for. Almost overnight, her breathing improved, her fever dropped, and her blood pressure was back to normal. "The ICU doctor who was on [duty] that night told us that he had never seen a patient that sick get better that quickly," Rheingold says.

Finding that tocilizumab could be an effective treatment for patients whose immune responses are too strong turned out to be a breakthrough discovery that has enabled CAR-Ts to fulfill their cancer-fighting potential. Scores of people have been saved because of what the CHOP medical team had learned from Emily's near-death experience.

Rheingold stated, "We were overjoyed when the results came back. Emily was in remission" three weeks later. "She had completely responded to her T-cell therapy. To see Emily go from leaving the hospital and starting to recover to basically going to school and playing soccer and looking like every other grade-school kid is just wonderful. I think this is the best reward I get from doing the work that I do with kids."

Now a thriving grade-school student, Emily has been in remission for four years and lives a happy, healthy life. However, it's still too early to call CTL019 a magic bullet for children with ALL.

"Half of sick children who normally would have been in hospice, are now being cured with CAR-T therapy," says Rheingold. "But our goal is to cure all of these children. They have such potential lives ready to be lived. We need to continue the research and perfect the technology so we can bring our cure rate to 100 percent. And then we have to expand the CAR-Ts so we can use them for other cancers that could benefit from this therapy."

NICHOLAS WILKINS' BATTLE WITH ALL

Acute lymphoblastic leukemia (ALL), a cancer of the blood and bone marrow, is the most common of all childhood cancers. Eighty-five percent of children with ALL are cured after a two-and-a-half-year process with standard chemotherapy. Fifteen percent, however, are not as lucky. Seventeen-year-old Nicholas Wilkins was first diagnosed with ALL in 2002, at just four years of age. At age seven, after undergoing standard treatment, he was in remission. He was happy, active in sports, and doing well at school.

When Nicholas was eleven, however, his medical condition changed. He began having stomach pains and when he played sports, he would quickly get winded. By a fortunate twist of fate, he injured his ankle playing football, which resulted in an ER visit. During the hospital visit, blood work revealed that his leukemia had returned. Soon he completed an intensive nineteen-week course of chemotherapy and radiation. The treatments triggered a wave of complications and challenges,

(continued on next page)

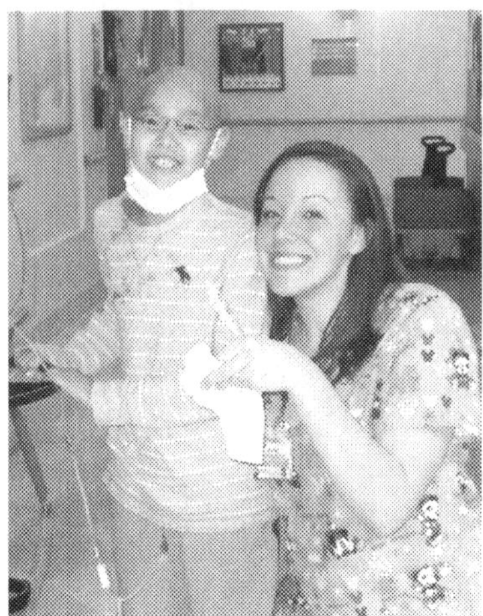

Left: Nick, a year after his first remission from leukemia, at his graduation from kindergarten in 2004.
Right: Nick during his treatment at Inova Fairfax Hospital in Virginia in the fall of 2009.

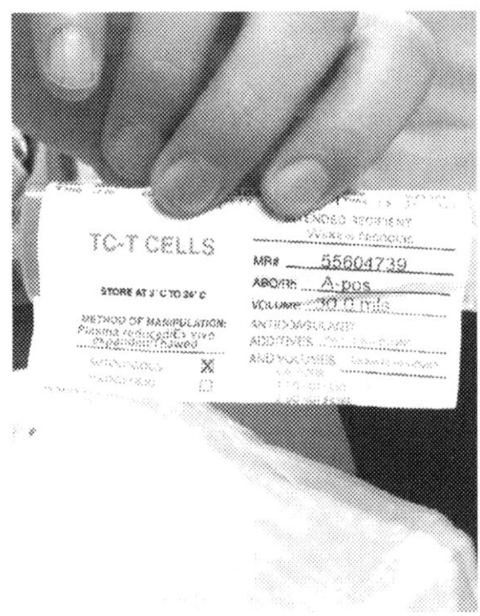

Left: Nick's modified T cells before reinfusion.
Right: Nick at the time of graduation from high school in 2016, three years after undergoing T-cell therapy.
(Photos courtesy of Lisa Wilkins)

including life-threatening staph infections and a burst appendix. Finally, on January 26, 2010, he received a bone marrow transplant from his older sister, Brittany. Again, Nicholas was able to return to school, apparently cancer free.

This remission was short lived, and, in 2013, his ALL returned. Even worse, an attempt to use traditional chemotherapy drugs failed. His cancer became resistant to drugs that had previously worked. Having exhausted all traditional ALL therapies, his family turned to clinical trials. They were referred to Dr. Susan Rheingold of the Pediatric Division of Oncology at the Children's Hospital of Philadelphia (CHOP). At CHOP, Dr. Rheingold took over his medical care and referred him to the CAR-T-19 clinical trial that was using chimeric antigen receptor T-cell therapy. In this revolutionary but experimental treatment, millions of his T cells were collected from his blood, then reengineered in a laboratory. On May 21, 2013, Nicholas was reinfused with the T cells so they could disperse to find and kill all cancerous cells. When the first test results came back, he was in remission. He had completely responded to his T-cell therapy. Three years, later he has no leukemia. The cancer-fighting T cells are still in his body, and he is cancer free, hopefully forever.

One limitation CAR-T cells have in eradicating cancer is that they themselves are sometimes attacked and eliminated by the immune system. To try to circumvent this weakness in CAR-T therapy, Rheingold and others are exploring the use of "armor" to protect the cells long enough for them to do their work. In what is perhaps an example of cellular payback, researchers are taking a page from cancer's own playbook. You will remember that cancer cells use the protein PD-1 to hide from the immune system. Researchers are using that same strategy, adding genes to CAR-T cells that tell them to make PD-1 or other proteins that have similar effects in helping medicinal CAR-T cells hide from the immune system. In 2016 a clinical trial of armored CAR-T cells against relapsed ovarian cancer started recruiting patients at Memorial Sloan Kettering Cancer Center.

IMMUNOTHERAPY BEYOND CANCER

Of course, immunotherapy isn't limited to cancer, and we will no doubt continue to find more applications in years to come. One interesting use is as a complement to bone marrow transplant in severe combined immunodeficiency (SCID), commonly called "bubble boy" disease. This condition results from an almost completely

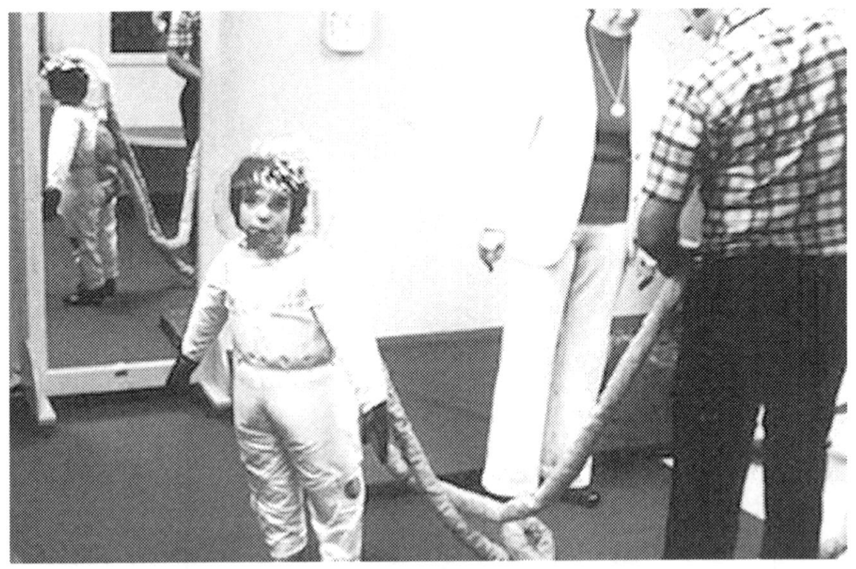

David Vetter was born September 21, 1971, with severe combined immunodeficiency (SCID) and, like Case Martin, spent his entire life in a sterile "bubble" to protect him from germs that, while innocuous to most people, could easily cause a fatal infection in SCID patients. Here David is in a special "spacesuit" built for him by NASA engineers that allowed him to venture outside his bubble for short distances. David died of Burkitt's lymphoma in 1984. (Public Image, NASA Johnson Space Center)

dysfunctional immune system, requiring that patients be isolated from any possible infection.

In 2011 six-month-old Case Martin couldn't fight off even the tiniest cold. At Texas Children's Hospital, doctors treated his condition with a stem cell transplant, hoping to replace Case's malfunctioning immune system with a new one. And to this standard treatment they added a special twist: to protect Case while his new immune system took hold, they gave him an infusion of about three million T cells engineered to attack dangerous viruses. The treatment worked, and now, five years later, Case's mom describes him as "happy, healthy, and full of energy." This immunotherapy is quickly becoming a standard addition to treatments that result in a severely impaired immune system.

Another iteration of this "adoptive T cell" strategy targets human immunodeficiency virus—HIV, the virus that causes AIDS. At the University of North Carolina, David Margolis, MD, has shown the technique is safe, and, in 2016, began trials combining adoptive T cells with the cancer drug vorinostat (ZOLINZA), hoping that this HDAC inhibitor will shock latent HIV viruses out of hiding so that his T cells can kill them.

Another kind of immunotherapy shows promise against type 1 diabetes. "In searching for an effective diabetes treatment, our current research goal is to replace insulin injections with nature's own solution," says David Pearce, PhD, president of Sanford Research, which is funded by the South Dakota businessman T. Denny Sanford. Recently, eighty-year-old Sanford committed his researchers to finding a cure for type 1 diabetes before he dies. "By taking advantage of a person's own immune system and, more specifically, by using their regulatory T cells, or Tregs, we hope to protect the beta cells in the pancreas from a wayward immune system attack," says Pearce. "These Tregs will also keep the immune system functioning properly so it can still fight off infections and cancers."

Tregs, which we first discussed in chapter seven, are in many ways the "good cop" counterbalance to the "bad cop" T effector cells that attack cells and tissues in autoimmune disease. When those T effector cells attack, Tregs tell them to stand down. But for some reason, in autoimmune disease, Tregs lose their ability to control the rogue T effector cells. Pearce hopes that an infusion of normal Tregs into a person with type 1 diabetes could help reeducate and rebalance the patient's immune system so that insulin-producing beta cells in the pancreas are protected from T-cell attack.

In 2015 the University of California, San Francisco, and Yale University finished a phase I trial with 14 patients, showing that infusing even up to 2.6 billion of the patient's own Tregs resulted in no serious side effects. Even in this preliminary trial, which was designed only to test for safety, 2 of the 14 patients were able to stop their insulin injections for up to two years after receiving this novel therapy.

Based on these promising results, the Sanford Project recently enrolled the first 18 patients with newly diagnosed type 1 diabetes into a study called T-Rex. This is an 111-patient phase II trial that will test

Dr. David Pearce president of Sanford Research. (Courtesy of Sanford Health)

Sanford Center in Sioux Falls, South Dakota. (Courtesy of Sanford Health)

the effectiveness of this Treg immunotherapy in providing long-term diabetes relief. In this study, children ages twelve to seventeen will have their blood drawn so that researchers can collect their Tregs. Back in the laboratory, these cells will be grown in a specially created nutrient mixture until they multiply by the millions. Days later, they will be infused back into the same patients they came from. Early data showed that not only are patients getting back many, many more Tregs, but the process of multiplying the original population of dysfunctional Tregs somehow "wakes them up." This vast number of reset Tregs is now able to control the T effector cells responsible for the autoimmune attack that had been destroying the insulin-producing beta cells in the pancreas.

If the Sanford Project's innovative Treg diabetes treatment is found to eliminate or significantly reduce the need for daily blood testing and insulin injections, a much larger and more definitive phase III study will soon follow.

"If using the patients' own Tregs . . . proves successful in the final T-Rex trial and leads to improved [native] insulin levels, Treg treatment could be available in several years for all patients with type 1 diabetes," says Pearce.

MOVING FORWARD WITH CELLULAR THERAPIES

For some forms of cancer, cell-based immunotherapy is a cure that is here today and ready for use. But for other kinds of cancer and for many other conditions, we are only at the dawn of the age of immunotherapy. Developing innovative medical strategies to put the cells of the immune system to work in other ways is destined to save even more lives. The question is not *if* this will happen, but when and how soon. The pace of discovery and testing is limited only by the ingenuity of researchers and the funding we can provide to test their inventions.

PART THREE

CHANGING OUR DNA

RARE DISEASES AND DESIGNER HUMANS

If we could make better humans, why shouldn't we?

—JAMES WATSON, PHD

CHAPTER NINE

THE GENE REPAIR TOOL KIT

We may be near the beginning of the end of genetic diseases.

—JENNIFER DOUDNA, PHD

Clara was an "easy" baby, engaged and happy. But by the time Clara was six months old, her mother, Laura, was worried that something was wrong. Sometimes when she was feeding Clara, Laura would look down into the baby's big, beautiful eyes and see that the pupils were shaking. When this happened, Clara never seemed to look back at her mother. The eye doctor saw the shaking, too—a symptom called nystagmus—but the exam was otherwise normal. Clara was developing as expected; aside from the shaking, her eyes seemed fine. But her mother knew something wasn't right.

When Clara's parents took her to Johns Hopkins Hospital in Baltimore for a second opinion, an MRI showed a slight Chiari malformation—a condition in which brain tissue extends into the spinal canal, restricting the flow of cerebrospinal fluid through the brain and spinal cord. Could that have been the cause of Clara's shaking eyes? It seemed plausible, but after surgery to remove a small section of bone at the base of her skull to improve the flow of fluid, the shaking remained.

Clara was now a year old, and when Laura turned on the light in her room, she didn't seem to notice. Clara was losing her sight, and no one knew why. Her parents turned to a nationally recognized eye specialist at Children's National Medical Center in Washington, DC. After an electroretinography to evaluate Clara's retina function, the doctor was pretty sure the child was suffering from a condition known as Leber congenital amaurosis (LCA). LCA is a degenerative disorder in which a faulty gene is unable to make a special form of vitamin A needed by the photoreceptor cells of the retina.

LCA affects about one in every 100,000 babies and is the most common cause of congenital blindness in children. It's a recessive autosomal genetic disorder, meaning you need two copies of a defective gene—one from each parent—in order to develop the condition. With only one bad copy, the normal gene—RPE65, which manufactures a repair protein—functions well enough to prevent disease, but that person is still a carrier of the mutation. But if both parents are carriers, each of their children has a one-in-four chance of inheriting two defective copies and certain blindness. Some children with LCA are born blind, while others progress into blindness over months or years. Clara seemed doomed to become completely blind by age five.

But if a faulty gene creates Clara's condition, why not just fix or replace the broken gene? If the gene is making something harmful, why not just turn it off? If the gene can't make something the body needs, just substitute a new gene that can.

That seems simple in concept, and in fact, the ability to turn on, turn off, or replace malfunctioning genes has been a major goal of biomedical science for at least half a century. Malfunctioning genes cause over a thousand recognized conditions, with many more diseases and disease subtypes being identified every year. The ability to edit genes and manipulate gene function would improve, lengthen, or save millions of lives every year. For example, cystic fibrosis is caused by inheriting two faulty copies of the gene CFTR. Huntington's disease is caused by mutated copies of the gene called huntingtin. Sickle cell disease is due to a slight change in the gene HgbS. Alterations in the gene BRCA make women much more likely to develop breast and ovarian cancer. (In fact, *all* cancers are caused by genetic changes, although most of these mutations are not inherited.) And HIV depends on a protein made by the gene CC5 in order for the virus to be able to enter cells. If we could only fix the broken bits of the human genome, we could prevent or cure these conditions. But as we'll discuss later in this chapter, messing with Mother Nature often has unintended consequences.

THE CRISPR REVOLUTION

Given the enormous promise of gene editing, scientists have tried to flip or repair genetic switches in a number of ways. One way is by using viruses to do the work. Almost by definition, a virus is a tiny machine made to insert genes into a host organism's genome. Generally, a virus inserts its *own* genes into human DNA, thereby tricking the cell's machinery into making more viruses. But scientists are using viruses loaded with "designer" genes as tools to paste new genetic code into sick or damaged cells. A 2009 letter to the *New England Journal of Medicine* reported some success with a viral-based gene therapy against LCA, the condition that is stealing the sight of Clara. But generally, viral-based gene therapies still need a great deal of development and fine-tuning. Where exactly will the virus insert the new gene? How often will the virus be successful?

So, in parallel with viral-based gene therapies, scientists have been developing ways to edit the genome directly. One way is with genomic scissors known as "zinc finger nucleases." These molecular scissors use engineered bits of DNA to find and snip out specific genetic targets, but they are extremely difficult to build and very expensive. Similar problems exist with several other gene-editing strategies. But a new technique based on a trick of the bacterial immune system has the potential to make all these other technologies obsolete.

CRISPR—which stands for clustered regularly interspaced short palindromic repeats—has revolutionized the way scientists work with genes

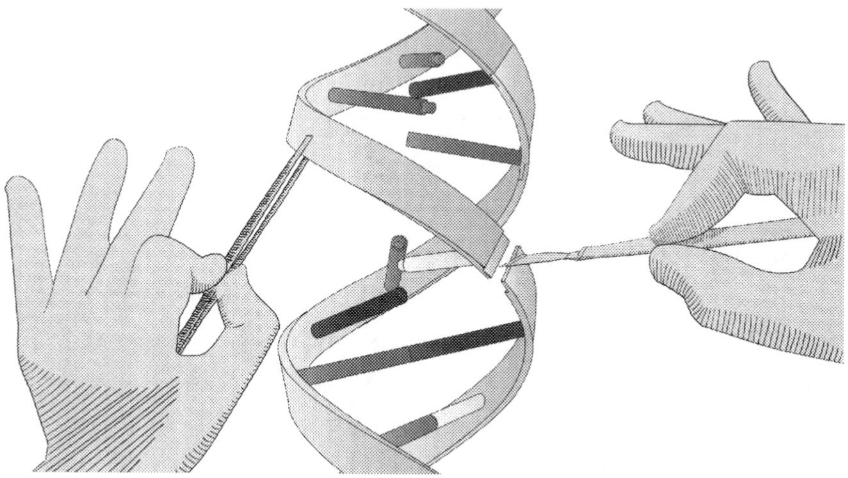

Scientists have developed increasingly precise methods to edit genes. The cutting and splicing of DNA is not done by hand, of course, but by molecular techniques.

in the lab and is now being tested by doctors on patients like Clara who suffer from genetic conditions. CRISPR works by appropriating a tool that bacteria use to recognize and destroy invading viruses. Basically, bacteria do this by splicing a bit of viral DNA into their own genome, packing it away like a collection of "WANTED" posters of criminals. It's a bit like a vaccine; inheriting this snippet of viral DNA increases the immunity of future generations of bacteria against viruses. However, it's the "attack" and not the "defense" side of this strategy that allows gene editing.

Here's how it works: As you know, DNA is the blueprint for proteins—but DNA doesn't communicate directly with the cell's protein-making machinery. Instead, the pattern of DNA is taken up by another player called RNA, which then moves through the cell to the place where proteins are made. It's the same in bacteria—a snippet of viral DNA leads to the production of matching RNA. But the bacteria added something special to this RNA, a Cas9 enzyme that can cleave. This enzyme is like a chain saw, able to cut completely through double-stranded DNA. The bacteria use this tandem—RNA with Cas9—to attack the virus the next time it appears. Since the RNA was built on viral DNA, when the RNA encounters *new* viral DNA, it positions itself next to it, and the Cas9 it's carrying saws that viral DNA in two at exactly the desired location.

The CRISPR-Cas9 combination can be used not only to cut DNA but to put new DNA in its place as well. Think about what this means. Scientists can now literally delete any gene from any genome *and* insert a new one in its place. We'll discuss the difficult ethics of this in a later chapter, but for now we'll focus on the promise.

In the lab, CRISPR is leading to a new understanding of disease. For example, take chronic myeloid leukemia (CML). Until the 1960s CML was a death sentence. CML is caused by an unfortunate trade of genetic material between chromosomes 9 and 22 that results in a "fusion" mutation known as the Philadelphia chromosome, because its discoverer, Peter C. Nowell, MD, found the abnormality in 1960 while working at the University of Pennsylvania's medical school.

The first step in medical and especially genetic research is often to develop an animal model with the disease or mutation in question so that researchers can experiment in ways not possible in humans. Starting in the 1960s that's exactly what Nowell and colleagues around the country tried to do. But getting the Philadelphia chromosome into mice was no easy task. First, they injected DNA synthesized to contain the Philadelphia chromosome into mouse embryos and hoped the change would be incorporated into the mouse genome. This rarely happened, and when it did, many mice grew into chimeras, in which some cells contained the Philadelphia chromosome while some contained normal DNA. Only some of these chimera mice passed on the Philadelphia chromosome to their offspring. It required multiple generations of breeding to create a mouse with the *two* copies of the Philadelphia chromosome needed to create the disease. This painstaking process took years, but finally, in the 1990s, researchers had an animal model of CML that allowed them to seek answers to crucial questions. The first was how to get rid of or neutralize the Philadelphia chromosome that causes CML.

The answer was the first widely successful targeted therapy against cancer, the drug imatinib mesylate (GLEEVEC), which was approved in the United States for treating CML in 2001. It has since been approved for treating other cancers, including acute lymphoblastic leukemia (ALL), first for adults and, in 2013, for children. This drug turns off the flow of energy that keeps the cancer cell dividing. In 2006 the *New England Journal of Medicine* published the five-year follow-up results of GLEEVEC against Philadelphia-positive CML. The disease was completely invisible in 98 percent of patients. GLEEVEC is a good example of how gene-directed therapy should work. Researchers discover the gene or genes responsible for a condition, other scientists create models of the disease with this genetic change, and then we discover and test treatments that repair, silence, or destroy this genetic cause.

Unfortunately, as in the case of CML and the Philadelphia chromosome, these models are terribly difficult to make. That is, they *used* to be difficult to make. With CRISPR, a guide RNA that matches the target gene is injected into a mouse embryo where it delivers Cas9 to cut through the DNA and either replace a faulty gene with a new sequence or at least leave the troublesome gene disabled. With CRISPR it takes half an hour and a few dollars to do what used to take years of work and a great deal of money.

Modeling genetic diseases is only the beginning of CRISPR's promise. For example, about 2 percent of children worldwide are allergic to eggs, or more specifically to a protein contained in egg whites. Their immune systems mistakenly see this protein as the mark of an invasion and release massive amounts of histamines to combat the threat, resulting in itching, inflammation, swelling, and other symptoms of an allergic reaction. It might sound like a small inconvenience for a child to avoid eggs, but a major complication of egg allergy is the inability to receive vaccines, many of which are produced using chicken eggs. Also, egg products are used in so many prepared foods and recipes that it's very difficult to avoid them completely.

But what if egg whites didn't have this allergenic protein? A March 2016 article in the journal *Nature* describes the work of Timothy Doran, PhD, a molecular biologist at the Commonwealth Scientific and Industrial Research Organisation in Geelong, Australia, who is using CRISPR to solve this problem in part because his own eleven-year-old daughter is severely allergic to eggs. In exactly the same way that a disease researcher might use CRISPR to erase or replace a gene linked to a condition, Doran uses CRISPR to hobble the gene so that chickens can't make the allergenic egg protein. If Doran's work succeeds, we may soon see the world's first hypoallergenic eggs.

Another study published in *Nature*, this one in April 2001, describes a genetically distinct population of honeybees that are especially "hive-proud." These proud bees are driven by their genes to keep their hives free of mites, infections, and fungi, so the hives survive even better than normal hives treated with expensive pesticides. In the 2001 study, University of Minnesota entomologists Marla Spivak, PhD, and Gary Reuter used artificial insemination and careful breeding to select for hive-proud genetics. Now, more than a decade later, San Francisco biotechnologist Brian Gillis is using CRISPR to cut and paste the genes that make hive-proud bees. Could CRISPR be the answer to the epidemic of honeybee colony–collapse disorder that every winter results in the death of 30 percent of beehives and costs many billions of dollars? Some researchers think so.

As you might imagine, food-crop geneticists are using CRISPR to engineer hardier, healthier, more drought- and disease-resistant strains of fruits and vegetables. This raises an obvious question: if we can change out a gene to make a better ear of corn, why not use CRISPR to make a healthier baby? And beyond that, if CRISPR could be used to repair a gene that causes cystic fibrosis or sickle cell disease, could it also be used to engineer a baby with a certain color of eyes or a higher IQ or a better chance of playing college football? In reality, even if the genes for these traits could be identified, it's unlikely that the code would lie in a single gene or only a few. Targeting all the necessary genes would likely be an enormous task. However, the possibility exists for scientists to play God and edit out any number of undesirable genetic traits.

CRISPR editing is not without its risks. The most obvious one is: What if a chicken needs the gene that makes the allergenic protein? That gene might be involved in other important functions. Many aspects of normal growth, development, and function are the result of complex and poorly understood interactions between many genes. Knocking out a problem gene could have far-reaching unintended consequences that might be deleterious to an organism.

There is also the risk of what scientists call off-target effects. You'll remember that CRISPR cuts DNA at any point that matches its target sequence. Let's say a CRISPR-Cas9 system was designed to snip DNA containing the code ACAAGATGCCA. It would snip this code not only in the target gene but in any other gene where that sequence appeared. Since there are nearly three billion base pairs in the human genome, there's a chance that we'll damage genes we're not targeting.

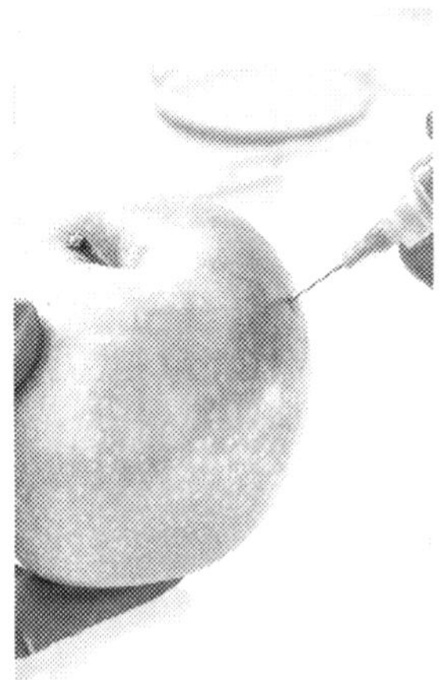

CRISPR can also be used to genetically modify foods (both plant and animal) much more efficiently than current methods. Genetically modified organisms (GMOs) can increase crop yield and be disease or insect resistant, among other commercially desirable characteristics, which is especially important for food crops such as wheat. The FDA has deemed most such crops safe for human consumption without the need to label them as GMO even though such labeling is required in more than sixty other countries.

This is what happened in the first-known instance of CRISPR being used to edit human DNA. In May 2015 the journal *Protein & Cell* published the results of a team of Chinese scientists who had used CRISPR to edit the genomes of nonviable embryos obtained from a fertility clinic. Their goal was to show how a CRISPR-Cas9 system could edit faulty copies of the gene HBB that lead to a class of blood disorders called beta thalassemia, a cause of anemia. The Chinese researchers had selected a seemingly attractive target; the disorder is caused by a mutation in a single gene, so fixing this gene should cure it—or so they thought. Unfortunately, the CRISPR-Cas9 system only worked in 28 of 54 tested embryos. And in those 28 embryos, the CRISPR-Cas9 system spliced the repaired HBB gene in several other matching sequences elsewhere in the embryo's genomes.

At present, the specter of these off-target effects is a major barrier to the widespread use of CRISPR in human clinical trials. But the first human clinical trial of CRISPR has taken a creative approach to solving the off-target problem—doing the gene editing outside the body. Funded by philanthropist and entrepreneur Sean Parker, the futuristic trial pairs CRISPR with immunotherapy to combat cancer. The trick is using CRISPR technology to engineer only immune system T cells, leaving the DNA of the patient's other cells unchanged.

As with many aspects of CRISPR, this gets a little complex, so it might help to review some immune system basics. One feature of the immune system is the ability of T cells to "learn" what to attack. An invading cell displays an antigen. If a T cell has a receptor on its surface that matches this antigen, the T cell attacks the foreign invader. A problem with cancer is that it's not necessarily marked with antigens that T cells naturally recognize. Since cancer cells develop from the body's own tissues, they carry many of the same markers as normal cells, so that T cells often fail to recognize them as bad guys. But what if we could engineer T cells that recognized the antigens that only cancer cells have?

Researchers at the University of Pennsylvania are now using CRISPR to grow T cells with exactly this feature. In this planned trial, doctors will remove cancer patients' T cells, "CRISPR" them, and then reinfuse them in the hope that they will recognize and attack the patients' cancers. The trial will be small, with only 18 people representing a range of tumor types. As an early stage clinical trial, a primary goal is to determine the safety of this CRISPR-based therapy, in addition to testing its efficacy. The trial is a milestone in realizing the promise of CRISPR for treating disease in human patients.

OTHER CRISPR CLINICAL TRIALS IN THE WORKS

Editas Medicine, a biotechnology company in Cambridge, Massachusetts, cofounded by MIT bioengineer and CRISPR coinventor Feng Zhang, PhD, was expected to begin human trials in 2017 to treat Leber congenital amaurosis, the disease that caused Clara's blindness.

Like the gene HBB in the Chinese embryo study, LCA is an attractive target: one gene, one disease. Also, the eye is easy to reach with genetic treatments, requiring only a simple injection. And the eye is a discreet compartment—what happens in the eye generally stays in the eye—so experimental treatments have less chance of producing system-wide damage. Finally, about 600 Americans have the form of LCA caused by this one faulty gene, making it a large enough population to recruit participants. In the Editas trial, a CRISPR-Cas9 system will target the gene responsible for LCA, snipping the defective copy and replacing it with a working one.

If successful, the Editas study will be the first to use CRISPR to adjust human DNA inside a human. It could also bring a ray of hope to children like Clara who are seeing their world darken around them. But Clara and other LCA sufferers represent only the beginning. With perseverance, creativity, passion, and maybe a touch of luck, CRISPR could augment or even replace traditional medicines for almost any genetic disease you can imagine.

CHAPTER TEN

REPAIRING DNA IN RARE DISEASES

An angel in the book of life wrote down my baby's birth,
Then whispered as she closed the book, "Too beautiful for earth."

—UNKNOWN

Each year four million babies are born in the United States. About 3 percent, or 120,000, will be diagnosed with a disorder that is caused by a mutation in just a single gene. For some of these conditions, there are treatments or medicines. For example, the drug ORKAMBI, FDA-approved in 2015, is designed to treat the roughly 8,500 cystic fibrosis patients in the United States who have a specific gene mutation. KANUMA, which was approved for the treatment of a metabolic disorder called lysosomal acid lipase deficiency, will be used for the twenty new patients expected to be diagnosed this year. But these are only two of more than 1,000 known genetic disorders, many of which affect only a handful of kids each year.

These are the "orphan" diseases, so called because each one affects so few people that they are often overlooked. Researchers tend to want to do the most good for the most people—such as the millions who die each year of heart disease and cancer—and it's a hard truth that pharmaceutical companies are unwilling to invest the many millions of dollars needed to bring a drug to market when that drug will reach only a few patients.

However, medicine is now rewriting the story of orphan diseases. This is partly due to increased visibility of these small patient communities that has been generated by passionate advocates. But another important factor is a new regulatory environment that is making it easier and more profitable to develop therapies for these kinds of diseases.

PUTTING THE SPOTLIGHT ON RARE DISEASES

"Rare diseases are among the most scientifically complex health challenges of our time," says Stephen C. Groft, PharmD, senior advisor at the National Center for Advancing Translational Sciences at the National Institutes of Health. "In the 1990s, most drug companies never would have considered developing a drug designed for 20 patients a year. But things are a lot different now because there are tremendous incentives to discover drugs that treat rare diseases."

Groft says that while approval of a new drug to treat a common cardiovascular condition might require a four-year study of 25,000 people, FDA approval for new drugs to treat orphan diseases can be accomplished with much smaller patient numbers, "with 200 patients or less," Groft says, in much shorter time.

Also, the market calculus for the development of new drugs targeting orphan diseases has shifted.

RARE DISEASES

The database of the National Organization for Rare Disorders (NORD) provides brief introductions for patients and their families to more than 1,200 rare diseases. This is not a comprehensive database since there are nearly 7,000 diseases considered rare in the United States. Here are brief descriptions of ten of these disorders.

1. **Aarskog syndrome** is an extremely rare genetic disorder marked by stunted growth that may not become obvious until the child is about three years of age, broad facial abnormalities, musculoskeletal and genital anomalies, and mild intellectual disability.

2. **Balo disease** is a rare and progressive variant of multiple sclerosis. It usually first appears in adulthood, but childhood cases have also been reported. While multiple sclerosis typically is a disease that waxes and wanes, Balo disease is different in that it tends to be rapidly progressive. The symptoms of Balo disease vary according to the areas of the brain that are affected. Symptoms may progress rapidly over several weeks or more slowly over two to three years.

3. **Canavan disease** is rare genetic neurological disorder characterized by the spongy degeneration of the white matter in the brain. Affected infants may appear normal at birth but usually develop symptoms between three to six months of age. Symptoms may include an abnormally large head (macrocephaly), lack of head control, severely diminished muscle tone resulting in "floppiness," and delays in reaching developmental milestones such as independent sitting and walking. Most affected children develop life-threatening complications by ten years of age.

4. **Dandy-Walker malformation (DWM)** is a brain malformation that occurs during embryonic development of the cerebellum and fourth ventricle. The cerebellum is the area of the brain that helps coordinate movement, and is also involved with cognition and behavior. The fourth ventricle is a fluid-filled space within the brain stem that channels fluid from inside the brain into the spinal cord. DWM is characterized by underdevelopment (small size and abnormal position) of the middle part of the cerebellum (cerebellar vermis), cystic enlargement of the fourth ventricle, and enlargement of the base of the skull (posterior fossa).

5. **Eales disease** is a rare disorder of sight that appears as an inflammation and white haze around the outer coat of the veins in the retina. The disorder is most prevalent among young males and normally affects both eyes. Usually, vision is suddenly blurred because the clear jelly that fills the eyeball behind the lens of the eye seeps out (vitreous hemorrhaging).

6. **Fanconi anemia (FA)** is a rare genetic disorder, in the category of inherited bone marrow failure syndromes. Half the patients are diagnosed before age ten, while about 10 percent are diagnosed as adults. Early diagnoses are facilitated in patients with birth defects, such as small size, abnormal thumbs and sometimes radial bones, skin pigmentation, small heads, small eyes, abnormal kidney structures, and cardiac and skeletal anomalies. The disorder is often associated with a progressive deficiency of all bone marrow production of blood cells, red blood cells, white blood cells, and platelets.

7. **Galactosemia** is a rare hereditary disorder of carbohydrate metabolism that affects the body's ability to convert galactose (a sugar contained in milk, including human mother's milk) to glucose (a different type of sugar). Classic galactosemia and clinical variant galactosemia can both result in life-threatening health problems unless treatment is started shortly after birth. A biochemical variant form of galactosemia termed Duarte is not thought to cause clinical disease.

8. **Hailey-Hailey disease** is a rare genetic disorder that is characterized by blisters and erosions most often affecting the neck, armpits, skin folds, and genitals. The lesions may come and go and usually heal without scarring. Sunlight, heat, sweating, and friction often aggravate the disorder. The symptoms of Hailey-Hailey disease occur because of the failure of skin cells to stick together, resulting in the breakdown of affected skin layers.

9. **Ivemark syndrome** is a rare disorder that affects multiple organ systems of the body. It is characterized by the absence (asplenia) or underdevelopment (hypoplasia) of the spleen, malformations of the heart, and the abnormal arrangement of the internal organs of the chest and abdomen. The symptoms of Ivemark syndrome can vary greatly depending upon the specific abnormalities present. Many infants have symptoms associated with abnormalities affecting the heart including bluish discoloration to the skin due to a lack of oxygen in the blood (cyanosis), heart murmurs, and signs of congestive heart failure. Ivemark syndrome often causes life-threatening complications during infancy. The exact cause of Ivemark syndrome is not known.

10. **Jarcho-Levin syndrome** is a rare genetic disorder characterized by distinctive malformations of bones of the spinal column (vertebrae) and ribs, respiratory insufficiency, and/or other abnormalities. Infants born with Jarcho-Levin syndrome have short necks, limited neck motion due to abnormalities of the cervical vertebrae, and short stature.

National Organization for Rare Disorders

"Despite the smaller patient pool for rare-disease research and development, several new economic factors are powering the development of orphan drugs. The government has provided tax credits and grants, waived FDA fees, and reduced timelines for clinical development, and there's a higher probability of regulatory approval. Commercial drivers include premium pricing, faster uptake, lower marketing costs, and longer market exclusivity," Groft says.

The National Institutes of Health now provide about $809 million in research funds annually in support of some 1,650 research projects specifically dealing with orphan diseases and drugs to treat them. In 2015 the FDA set a record for approvals of new drugs to treat rare diseases, sanctioning twenty-one new "orphan" drugs, which represent a surprising 47 percent of all novel new drugs approved that year. It was the second consecutive year in which the FDA set such a record.

"These factors combine to make this a golden age for new orphan disease treatments," Groft says.

But while awareness and economics have changed the game for orphan drug development, a paradigm shift in the way researchers categorize these conditions has the potential to change it in even greater ways. While a single orphan disease affects a small population, taken together, these 1,000 conditions significantly impact the lives of millions of people in North America alone. Today, more than 350 million people around the world have some kind of rare disorder, ranging from relatively well-known ailments such as cystic fibrosis and multiple myeloma to obscure, ultra-rare conditions such as dancing-eyes-dancing-feet syndrome, fibrodysplasia ossificans progressiva, lysosomal acid lipase deficiency, platyspondylic lethal skeletal dysplasia Torrance type, and Swyer-James syndrome.

What if, instead of looking at each of these rare genetic conditions as a unique, isolated challenge, we saw the entirety of rare genetic diseases as one problem with a common cause? Though the individual gene or genes that are broken in these conditions vary, the fact remains that 1,000 of these conditions are caused by the malfunction of a single gene. What if instead of focusing on which gene needs fixing, we tried to develop the ability to fix single genes? These 1,000 orphan diseases would become one condition—a sort of umbrella "single-gene malfunction" ailment. If we did that, this one ailment might have one cure—the cure mothers like Pat Furlong have been praying for.

FIXING THE GENE OF DUCHENNE MUSCULAR DYSTROPHY

"It was the mom in me. That's how I knew," says Furlong. In the early 1980s, this former nurse was concerned about her two infant sons, Christopher and Patrick. Christopher, the older by two years, was late to walk, while Patrick had trouble standing. When Furlong lifted Patrick up onto his feet, he would quickly slip back down to the floor.

Pat Furlong, founder of the Parent Project Muscular Dystrophy, a not-for-profit organization helping families of children affected by Duchenne muscular dystrophy. (Courtesy of Pat Furlong and Parent Project Muscular Dystrophy, EndDuchenne.org)

Her two older daughters had no such problems. "The boys had trouble walking up the stairs, while this posed no problem for the girls at a similar age," she says.

Though both boys were intelligent and outgoing, Furlong felt in her heart that something was not right. However, doctors, including her husband, Tom, a family practitioner, told her nothing was wrong with the boys. But when Chris was six years old and started to limp after riding his tricycle and complained of unexplained calf muscle pain, an orthopedic surgeon in their hometown of Middletown, Ohio, quickly confirmed that Chris had not suffered a muscle tear. It was something worse: Duchenne muscular dystrophy, a rare muscle-wasting disease that affects males. As the muscle tissues die, they are replaced by scar tissue, which is why Chris' calf muscles were so oversized for his age.

Three days of testing at Cincinnati Children's Hospital confirmed that both boys had Duchenne. Furlong was told that the disease typically began slowly in early childhood and moved on to progressive loss of muscle strength. Her sons would spend their teenage years in wheelchairs and would probably die of respiratory failure before the age of twenty.

Duchenne is truly a rare disease. First described in 1861 by Dr. Guillaume Duchenne, a French neurologist, the disease occurs in one in every 3,600 male births. There are about 15,000 boys living with Duchenne in the United States. The condition is caused by a malfunction of the gene that makes dystrophin, a protein involved in maintaining the integrity of muscle.

This gene exists only on the X chromosome. Boys with Duchenne symptoms have inherited from their mothers a faulty copy of the gene on an X chromosome. Because females have two X chromosomes, the normal, functional gene on the second X chromosome cancels out the defective one, so girls can unknowingly carry the faulty gene without getting sick. But since males have only one X chromosome (the other, inherited from the father, is a Y chromosome), they have no second, functional copy of the gene. (In scientific terms, the defective gene is

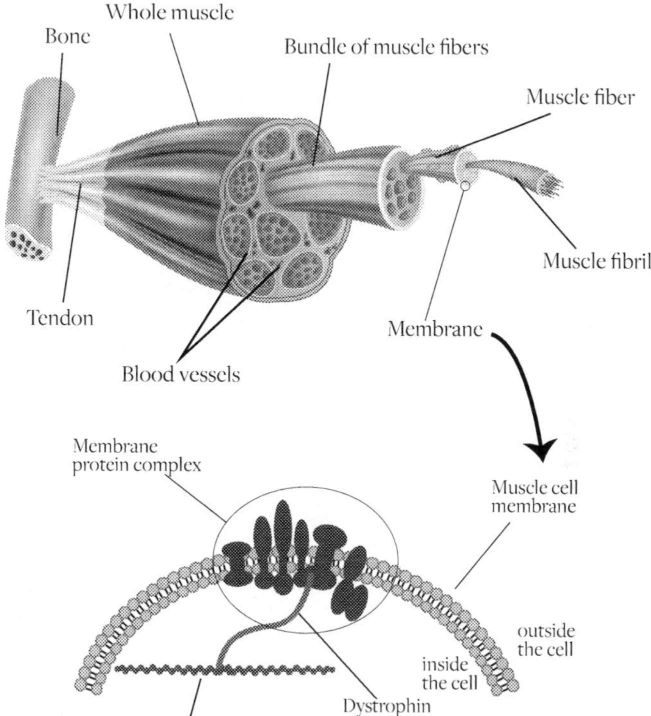

Duchenne muscular dystrophy is caused by a mutation in the gene that produces an important muscle protein called dystrophin, which is not produced. Muscles are made up of bundles of fibers (cells). A group of interdependent proteins along the membrane surrounding each fiber helps to keep muscle cells working properly. When one of these proteins, dystrophin, is absent, the result is Duchenne muscular dystrophy.

X-linked recessive.) The onset of symptoms usually occurs between the ages of three and five. Most boys are unable to walk by age twelve and soon thereafter need a respirator to breathe. Because there are so few patients, little research had been directed at the disease, and there were no treatments.

In 1984, when Pat Furlong's sons were diagnosed, her doctor advised her to take the boys home and love them, because they were going to die. "The doctor said that our lives were going to be blown up, that my two daughters were not going to like me as much, because I would be giving so much attention to my sons," she says.

"The day of that diagnosis and many times thereafter, I thought 'Let's all jump off a bridge and go to heaven together.'" Instead, Furlong decided to fight.

Furlong borrowed $100,000 from a local bank and used it to visit leading medical centers in the United States and Europe. She met with doctors and researchers who were looking into the cause and possible therapies for Duchenne. She helped provide funding to launch a small clinical trial of a compound, adenylosuccinic acid, and enrolled her sons in the study. Two years later, their condition worsening, Furlong took them out of the study. Another trial involving immature muscle cells called myoblasts soon followed, but this therapy didn't work either.

By 1994 there were no more drug trials to enter. As predicted, Christopher and Patrick soon lost the use of their muscles and legs. They needed help with eating, toileting, bathing, getting in and out of their motorized wheelchairs, and getting into bed. In 1995, when Christopher was 17, he developed a cold that turned to pneumonia. These were hard times for Furlong. She cried every day. "My son Christopher once

Muscular dystrophy causes progressive skeletal muscle deterioration, eventually requiring children to use wheelchairs for mobility.

said to me, when he saw tears in my eyes, 'What are you crying about?' I said to him, 'I just need a miracle for you and your brother.'"

"Is it fair that there is only a miracle for one or two boys instead of all boys?" Christopher replied.

Christopher died several weeks later. Patrick developed a similar respiratory problem seven months later and died at the age of fifteen. Never forgetting Christopher's words, Pat kept fighting even after her sons passed. She started the group Parent Project Muscular Dystrophy and has been at the forefront of searching for the "miracle" that could help all boys with Duchenne. Increasingly, Furlong sees the potential for this miracle in gene editing, specifically in the revolutionary CRISPR-Cas9 gene-editing system, which we discussed in chapter nine.

"Gene editing takes advantage of systems identified in other organisms in order to correct 'typos' in the genetic code," Furlong says. In the case of Duchenne, the typo is a malfunctioning copy of the gene Xp21, which codes for dystrophin, the aforementioned muscle-repair protein. Theoretically, CRISPR is the ideal way to fix the condition. Furlong lists three ways CRISPR could be used in Duchenne: "The first is editing of the dystrophin gene in germ cells [sperm and eggs] or in embryos. The second is editing of the dystrophin gene outside the body, in muscle stem cells or satellite cells that are then transferred back into the patient to populate muscles with these 'engineered' cells that are capable of producing dystrophin. Finally, there would be the introduction of CRISPR-Cas9 reagents directly into the muscles of the Duchenne patient, to edit the dystrophin mutations inside the body."

All three strategies are promising, and all three come with challenges. Let's look at her suggestions individually.

First, editing the gene in embryos (or even in mothers who find that they are carriers) would forever alter the child's genome and the genome of any future generations fathered by that child. On the surface, this sounds desirable. Who wouldn't want to eliminate the Duchenne gene from the population? But pursuing this compassionate and controlled use of CRISPR requires opening the same door as any other kind of human genetic engineering and dealing with the sticky ethical questions that lie inside. If we allow the use of CRISPR to modify the germ lines of babies affected with Duchenne, where then do we draw the line? (We'll discuss this in depth in chapter twelve.)

Furlong's third suggestion, to introduce CRISPR technology inside the muscles of those affected by Duchenne, is less ethically challenging, but brings with it significant risks to the patient. Because CRISPR reads the code of genes and then snips the genome every place it recognizes its target code, it may accidentally snip DNA anywhere the target code sequence appears throughout the entire genome. The human genome, or genetic code, is like a book written in a language whose alphabet has only four letters, A, T, G, and C (which stand for the nucleotides adenine, thymine, guanine, and cytosine). However, it's a very long book, containing a total of about three billion pairs of these four "letters," which make up a total of about 20,000 genes. So chance dictates that many small sequences of code are repeated in the makeup of different genes. What if CRISPR accidentally cuts a gene unrelated to Duchenne but essential to something else? This problem is a barrier to using CRISPR against many diseases, but many Duchenne boys and their parents may reason that such a risk is worth taking when the alternative is certain death.

Furlong's second suggestion, to harvest muscle cells or muscle stem cells, engineer them with CRISPR outside the body, and then reinfuse these cells into the patient, splits the difference between the two other ideas. It seems relatively safe and doesn't affect the genome in a heritable way—but its effectiveness depends on, among other challenges, the ability of these engineered cells to stitch themselves back into the fabric of a patient's muscles in a way that lets them survive long enough to deliver dystrophin in enough quantity to affect the condition.

With support from Furlong's organization and others, researchers are moving forward with all three strategies. For example, in the laboratory of Eric N. Olson, PhD, at the University of Texas Southwestern, his team has used CRISPR in the germ cells of mice to correct the Duchenne mutation, showing that if only 40 percent of genes are corrected, animals go on to develop normally. Collaborating with universities around the world, Charles Gersbach, PhD, of Duke University, Ronald Cohn, MD, of the Hospital for Sick Children in Toronto, and Akitsu Hotta, PhD, of Kyoto University have successfully edited cells isolated from Duchenne patients outside the body that could theoretically be reinfused to boost the production of dystrophin. And the laboratory of Renzhi Han, PhD, at the Ohio State University has corrected the Duchenne gene in the cells of mouse muscles by using electrical current to help CRISPR reagents reach deep into the tissue.

Of course, mice are not men, so there remain technical, ethical, and safety hurdles before CRISPR can be used in any of these ways with human Duchenne patients. Researchers are currently exploring issues from the design of the RNA guide molecules to better focus the specificity of DNA targeted by CRISPR, as well as the best systems for delivering CRISPR components into muscles. The immune system may also be a challenge. Will it attack and render useless CRISPR reagents or cells engineered outside the body and then reinfused? Will the immune system activate against newly made dystrophin itself? That said, while CRISPR-based treatments for Duchenne are still confined to the laboratory, other conditions are already green-lighted for human trials.

CRISPR IN HUMAN TRIALS

Feng Zhang, PhD, whom we noted in chapter nine as one of the bioengineers essential to the invention of CRISPR, expects to begin human trials in 2017 using CRISPR to treat a rare eye disease called Leber congenital amaurosis (LCA), also discussed in chapter nine in the case of the child Clara. Affecting about one in every 80,000 infants, the condition is usually diagnosed early in life and quickly leads to complete blindness.

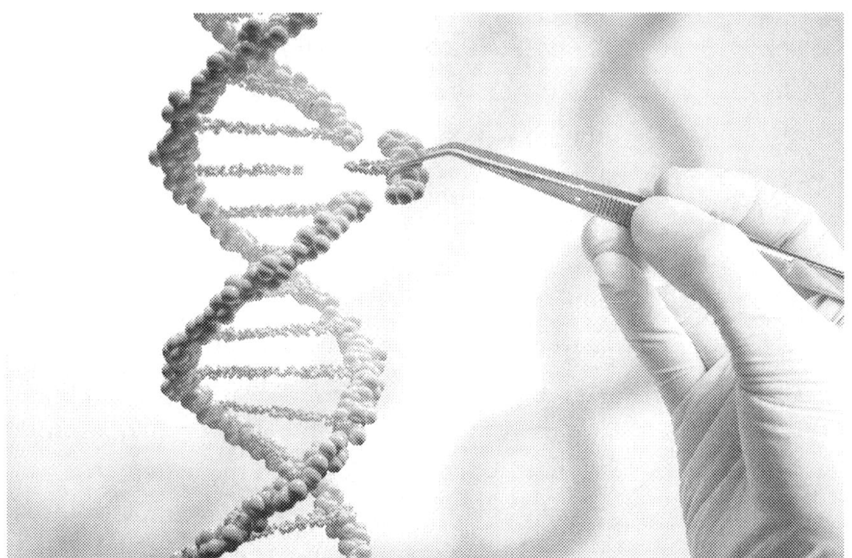

This is a nearly ideal condition on which to test CRISPR inside a patient's body for several reasons. The eye is an attractive test site because cells engineered by CRISPR can't travel from the compartment of the eye to other areas of the body; any potential danger starts and stops inside an eye that is or soon will be blind anyway. Moreover, the subset of LCA that will be the subject of this trial is caused by a single malfunctioning gene; fix this one gene, and you've fixed the problem. Also, the eye is easy to reach with treatments. And finally, LCA has already been the test disease for the first experimental gene therapies using strategies other than CRISPR. In 2008 doctors used a viral delivery system to splice healthy copies of the gene RPE65 into the DNA of eye cells in people suffering from a variety of LCA related to that gene. The journals *Human Gene Therapy* and *Molecular Therapy* both reported that a year and a half later many patients in the trial regained sight and that there had been very few side effects. With successful gene therapy already directed at LCA, it's not a giant leap to accomplish the same genetic insertion via a new technology.

In fact, the trial conducted by Zhang's Cambridge, Massachusetts, company Editas will use viruses as well, but instead of using them directly to insert a healthy gene into malfunctioning eye cells, these viruses will manufacture the components of a CRISPR system. Once assembled inside the eye, the CRISPR-Cas9 system will delete about 1,000 DNA letters from the CEP290 gene in the patient's photoreceptor cells and replace these with 1,000 new letters. Laboratory experiments leading up to the approval of the trial showed that the gene should function correctly again after the CRISPR edit.

Even sooner, the University of Pennsylvania will oversee a trial that uses CRISPR outside the body to engineer a patient's T cells to attack cancer. This trial (which is not against a rare disease, but has implications for the field) puts a CRISPR spin on the CAR-T T-cell therapy profiled in chapter five and on cancer immunotherapy. In this technique, T cells harvested from a patient's blood are engineered to recognize specific proteins that mark the surfaces of cancer cells and are then reinfused into the patient's body to find and fight the disease. In the Penn trial, CRISPR will be the mechanism of this *ex vivo* (outside the body) engineering.

Specifically, led by immunotherapy pioneer Carl June, MD, the group will use tried-and-true techniques (targeting the PD-1 gene as described in chapter five) to add a protein to T cells that helps them find cancer, and then also use CRISPR to knock out a second gene that helps cancer cells deactivate T cells. The two-year trial will treat 18 patients with myeloma, sarcoma, or melanoma who have stopped responding to approved therapies.

These specific approaches are among the few CRISPR therapies that are in or near human clinical trials so far. But on their heels are dozens if not hundreds of CRISPR-based systems in development at the many biotechs that have sprung up to capitalize on this elegant ability to cut and paste pieces of the genome. For example, alongside Editas in Cambridge is the company CRISPR Therapeutics, founded by another codiscoverer of the CRISPR technology, Emmanuelle Charpentier, PhD. The company lists in its pipeline CRISPR-based medicines for beta thalassemia, sickle cell disease, Hurler syndrome, severe combined immunodeficiency (SCID), glycogen storage disease, hemophilia, cystic fibrosis, and, last but certainly not least for Pat Furlong, Duchenne muscular dystrophy. With financial incentives now available for the development of medicines to target rare diseases, regulatory challenges smoothed by a new era of expedited FDA rules for orphan conditions, and now the ability to edit specific genes with CRISPR, this company and others are poised to start checking off rare diseases one by one.

Pat Furlong's son Christopher asked why the miracle that his mom sought for his condition should be limited to him and his brother, spurring Furlong to crusade for a cure on behalf of the entire community affected by Duchenne. A similar theme runs through the entire field of orphan diseases—in the sense that the miracle of CRISPR-Cas9 gene editing might not only cure Duchenne or a rare eye disorder or make cancer immunotherapies more effective, but also it might fix broken genes for virtually any of a thousand conditions caused by the mutation of a single gene.

Despite technical and ethical hurdles, the promise of CRISPR remains immense. And now, with the recent start of human trials, we are starting to realize the promise of curing rare diseases by editing the genes that cause them. Pat Furlong's miracle is becoming reality.

CHAPTER ELEVEN

PUTTING A BULL'S-EYE ON SICK CELLS

Float like a butterfly, sting like a bee. The hands can't hit what the eyes can't see.

—MUHAMMAD ALI

As you read this challenging chapter, keep in mind one important fact: the cells of nearly every disease are genetically different from the healthy cells in our bodies, even if those differences are sometimes very slight. As you've seen, the gene-editing tool CRISPR is being used to fix some of these aberrations. In diseases or conditions caused by single-gene mutations, CRISPR can snip out the bad gene and replace it with a good gene. However, for many diseases, the problem is far more complex. For example, in certain cancers, sickle cell disease, infectious diseases like Zika and West Nile, or staph infections, there are many, many differences between the cells of these diseases—or, in the case of viruses, cells *infected* by these diseases—and your healthy cells. No simple genetic swap can fix all the problems.

So doctors are still limited to taking on these cells the old-fashioned way—by trying to kill them. Unfortunately, the old-fashioned way tends to be extremely imprecise. We are only recently learning how to use differences in cell morphology to selectively target them, hoping to replace chemotherapies doctors have depended on that do almost as much harm as good. For example, they might exploit the fact that cancer cells divide faster than normal cells or that they have difficulty repairing DNA damage. But this has been like gillnetting Pacific salmon. Although net makers can shape the holes to the size of the fish they want to catch, there are many more creatures of similar size in the ocean that will be caught in the nets.

So too with many medicines. Despite refinements in their use, chemotherapies still kill millions of healthy cells, and antibiotics decimate the good bacteria of your gut microbiome along with invading bugs. Not only does the indiscriminate and imprecise nature of many of our medicines lead to uncomfortable side effects, but these side effects often limit their use. Owing to accumulating mutations, weakening immune systems, and often years of exposure to environmental toxins (among other factors), many of our most dangerous diseases are found primarily in older patients or in those whose systems are already challenged by other conditions. Comorbidities (as doctors call them), age, and even progression of the original disease make many of our medicines unusable, as some of these therapies kill fragile patients more surely or quickly than the condition itself.

For patients in otherwise robust health, a dangerous medicine that poisons cells, based on an unclear understanding of what makes a disease cell different from a healthy one, is usually better than no treatment at all. But unfortunately, "no treatment" is the reality for many of our most dangerous conditions. There is no real treatment for Alzheimer's disease. There is no treatment for Zika. There are no good options for many cancers. Despite billions of research dollars, there is no cure for HIV. And even things like bacterial infections that seemed to have been put to sleep in the twentieth century with the development of antibiotics continue to vex modern medicine by evolving in ways that make them resistant to antibiotics.

If we knew the differences between healthy cells and those that are genetically diseased (like cancer) or infected with disease (like Zika), we could cure these ailments. Today, that's not just wishful thinking. In fact, this strategy of spotting and exploiting genetic differences forms the backbone of the modern shift toward personalized medicine. Take, for example, the drug imatinib (GLEEVEC).

In 1960 Peter C. Nowell, MD, of the University of Pennsylvania and David A. Hungerford, PhD, of the Fox Chase Cancer Center in Philadelphia reported that they had found shortened chromosomes in patients with chronic myeloid leukemia (CML). For decades this "Philadelphia chromosome," as it became known, was no more than an interesting footnote to the condition. The fact that people with CML had shortened copies of chromosome 22 didn't seem nearly as important as the fact that they almost all died within five years of diagnosis. Then, in 1973, scientists at the University of Chicago discovered what had shortened this chromosome. It turned out that in these patients, both chromosome 9 and chromosome 22 had broken, and when the body's DNA-repair mechanism tried to fix the damage, it stitched the halves together incorrectly, swapping part of chromosome 9 with part of 22. Chromosome 22 got the short end of the trade, and where it was now joined to a piece of chromosome 9, a new gene was formed—the fusion gene BCR-ABL. This new fusion gene was more than an interesting artifact of the leukemia—it was actually driving the cancer. Specifically, BCR-ABL was making a new enzyme—a tyrosine kinase—that was whipping white blood cells into uncontrolled, cancerous replication.

At Oregon Health & Science University, Brian Druker, MD, specialized in the study of tyrosine kinases, enzymes that help provide the energy for a cell's internal machinery. He began screening tyrosine kinase inhibitors (TKI) against samples of CML. One compound,

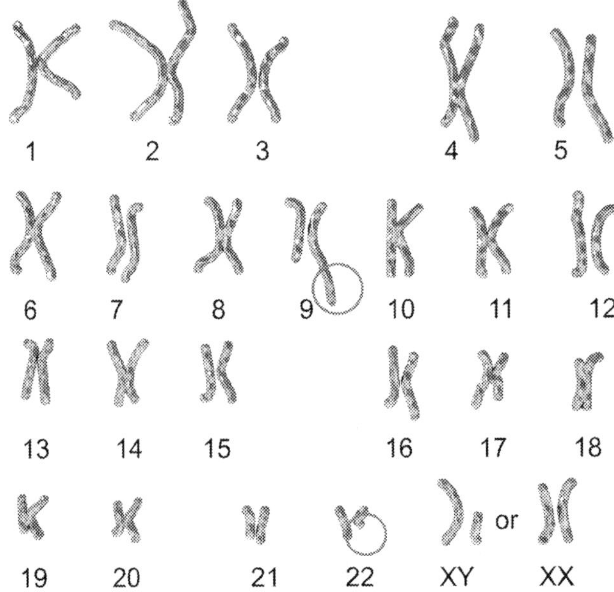

The 23 pairs of human chromosomes showing the portions on chromosomes 9 and 22 that are swapped (technically called a translocation) and fused to form a mutation called the Philadelphia chromosome.

then known as STI571, killed every CML cell in every dish in which it was tested. In 1998 Druker started testing the new drug, now named imatinib, in patients. The results were stunning. All 31 CML patients who initially tried the drug saw their blood counts return to normal. Like a switch, BCR-ABL turned on CML, and like a switch, imatinib turned it off. Since FDA approval in 2001, the drug has been marketed as GLEEVEC and has been the major factor in establishing a cure rate for CML of more than 95 percent.

The success of GLEEVEC shifted the bedrock of cancer research. If the Philadelphia chromosome caused CML and could be indirectly muted by the TKI GLEEVEC, then curing cancer might be as simple as pairing cancer-causing genes (oncogenes) with inhibitors that block what the gene makes. Sometimes this story line has been easy to repeat. For example, in 2007 a group of Japanese researchers reported that another fusion protein was driving a subset of lung cancers. In these patients, who were mostly nonsmokers, rotten luck had led to the gene ALK fusing with the gene EML4. Now, instead of one gene making an ALK protein and another making an EML4 protein, the mash-up of these two was making something entirely new—and this new fusion protein was causing cancer.

Fortunately, there was a small-molecule inhibitor already being tested in the clinic that inhibited ALK. The drug PF-02341066 was developed to silence another genetic player thought to cause cancer, but it worked even better against ALK. Just four months after ALK-EML4 fusion was first described as an oncogene, the first patient was treated with an ALK inhibitor. The drug, now known as crizotinib (XALKORI), led to almost immediate dramatic improvement.

Other successes have followed this model. Cetuximab (ERBITUX) targets the gene EGFR and is used primarily to treat colorectal cancers; ipilimumab (YERVOY) targets CTLA-4 and is used primarily against melanoma; bevacizumab (Avastin), which targets VEG-F, is used primarily in colon cancers; nivolumab (OPDIVO) targets PD-1 and is also mostly used to treat melanoma; olaparib (LYNPARZA) targets PARP, which causes some ovarian and breast cancers; vemurafenib (ZELBORAF) targets BRAF and is used to treat melanoma; and the list goes on.

Of course, not every pass through this cycle of discovery has led to cures as dramatic as GLEEVEC for CML. CML is special—there is only one cause, and when that avenue is blocked, the disease is through. Lung cancer, for example, is different—there are many mutations and translocations and other genetic alterations that can drive the condition. So when doctors use crizotinib (XALKORI) to block a cancer's ALK-EML4 dependence, lung cancers almost inevitably evolve (or already have evolved) another genetic addiction that allows them to work around the medicine's blockade. That said, crizotinib remains a powerful drug, as are many of the other tyrosine kinase inhibitors. And despite varying effects, the story line of development remains the same: find the genetic difference between diseased and healthy cells and then find or design a drug to exploit it.

You would think that modern gene sequencing would make this first step easy—just line up the three billion base pairs of a cancer cell against the three billion base pairs of a healthy cell and look for the points where the letters misalign. But like many things in medical science, this seemingly simple story line is anything but. When you line up a cancer cell with a healthy cell, you find thousands of differences, most of them meaningless. How can you tell which changes are meaningful and which are just harmless variations or random mutations that are not useful drug targets?

Finally, we are back to CRISPR and to this chapter's title. Only, the title is a bit misleading—it's not necessarily that scientists use CRISPR to paint bull's-eyes on diseased cells, but that they are learning how to use CRISPR to see the bull's-eyes that have been hidden there all along.

Here's how it works. CRISPR can be used to snip a single gene from a single diseased cell. But it can also be used to snip a single gene from 20,000 diseased cells. Scientists are now using CRISPR to snip a *different* gene from each of these cells in turn. When they do, some snips result in the death of these disease cells. By this method, scientists can ask which gene is required for the survival of this disease. When they find genes required by the disease but not required by healthy cells, then they've found a drug target.

That's what a team from the Wellcome Trust Sanger Institute did with acute myeloid leukemia (AML) cells. They used CRISPR to knock out a different single gene in each of 20,000 isolated populations of cells. Five hundred of these knockouts resulted in the death of the cells. Of those genes, 200 made products against which drugs could conceivably be developed. These 200 included some previously known oncogenes like DOT1L, BCL2, and MEN1, but most of the others were new discoveries. And a very few of these resulted in the death of AML cells *without* also killing healthy cells. But one stood out from the others: the gene KAT2A. When they used CRISPR to eliminate this gene in AML cells, they died. When the researchers got rid of KAT2A in healthy cells, it had no appreciable effect. Mice engineered to develop AML lived longer when researchers turned off this gene.

"KAT2A inhibition now needs to be investigated as a treatment strategy for acute myeloid leukemia," says George Vassiliou, PhD, joint project leader from the Sanger Institute and consultant hematologist at Cambridge University Hospitals NHS Foundation Trust. "Our hope is that this work will lead to more effective treatments against AML that will improve both the survival and the quality of life of patients."

A study in the May 2015 issue of the journal *Nature Biotechnology* describes another way to keep dangerous genes inactive. One way to keep a dog from biting you is to put a stick in its mouth, and a similar strategy works against genes. That's because many proteins have a "binding pocket" that needs to trap another molecule (often energy in the form of ATP, cells' basic unit of energy) in order to become active. Plug that binding pocket or otherwise disrupt it, and the protein stays deactivated. Researchers from Cold Spring Harbor Laboratory on Long Island used CRISPR to completely edit out these binding pockets. Again, they did this sequentially, deactivating many thousands of pockets until they found the ones that silenced proteins that a given disease needs to survive.

You were promised a bull's-eye, and there it is. Gene editing can now show us at the level of a protein's molecular structure how to reach diseased cells with new drugs.

This approach is useful far beyond cancer.

ZIKA: DEALING WITH A GLOBAL EMERGENCY

The Zika virus was first identified in rhesus monkeys in 1947 and was confirmed to have spread to humans five years later. For whatever reason, transmission of the virus remained rare through the second half of the twentieth century, until, as reported in an article in the *New England Journal of Medicine,* there was an outbreak in 2007 on the Island of Yap in the Federated States of Micronesia. Blood sample testing at a Centers for Disease Control facility in Fort Collins, Colorado,

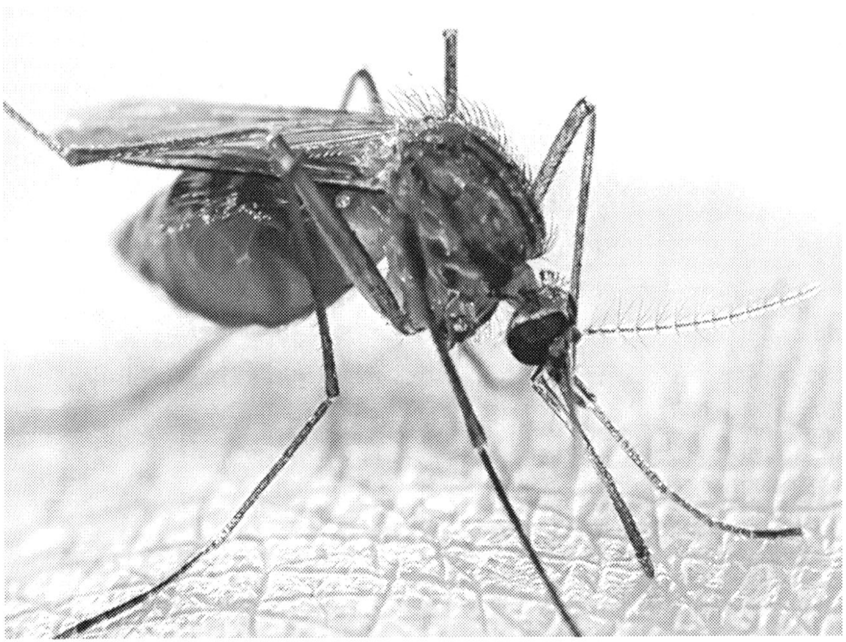

The Zika virus is transmitted through the bites of infected mosquitoes.

showed that about 14 percent of the island's 7,391 people had contracted the virus. By 2016 Zika infections in humans—the disease is transmitted by mosquitoes—had been confirmed in all of the Americas with the exception of Canada, Chile, and Uruguay. By March 2016 Brazil had confirmed more than 91,000 cases.

This disease inflicts harm in a particularly insidious way. In adults, the worst symptom is generally a mild rash, and many of those infected with Zika show no symptoms at all. The real damage Zika inflicts is to the babies born to women who often have no idea they are infected. Researchers are still trying to nail down credible numbers, but it is clear that a baby born to a Zika-infected mother has a significantly increased chance of birth defects, especially microcephaly, in which the brain and skull fail to grow and develop normally.

While there may soon be a vaccine to prevent Zika transmission, there is no treatment for infection. In the hunt for one, researchers from Washington University School of Medicine in Saint Louis used the

CRISPR-screen procedure against this infectious disease. Their results are reported in the November 2015 issue of the journal *Nature*. In this case, instead of asking which genes Zika needs to survive—because it's unlikely they could track down and treat every mosquito carrying the virus—they instead asked which genes must be present in humans for the virus to succeed with infection. As in screening cancer cells, they used CRISPR to turn off the genes in human cells, each in turn, and then challenged cells with Zika.

Most cells became infected, but a few did not. In these uninfected cells, it was genes in a family called SPCS that had been turned off. Zika "capsules" were still present in these cells, but they had been unable to split open and eject virus particles that would then burrow into host DNA, hijacking the cell's manufacturing mechanisms to produce more viruses. These SPCS genes are now one strategy being pursued in the race for a Zika vaccine.

A study in the July 2015 issue of *Cell Reports* describes a similar strategy against the West Nile virus, which is related to Zika. The goal, as with Zika, was not to attack the virus directly but to keep it from killing brain cells. To do this, researchers used a library of 77,406 CRISPR systems to sequentially delete 20,121 genes. When CRISPR deleted any one of seven genes—EMC2, EMC3, SEL1L, DERL2, UBE2G2, UBE2J1, or HRD1—West Nile lost its ability to kill cultures of human brain cells.

There's a very interesting and even more important takeaway from this research: all seven of these genes are involved in deciding how quickly proteins naturally decay. And so it seems as if West Nile depends on using these genes to accelerate the pace of protein decay to kill brain cells. Perhaps in this case, rather than showing us a specific genetic target, CRISPR shows us the biological system in which medicines must intercede. Drugs that protect the natural pace of protein decay may be as successful as drugs that target the genetics.

USING CRISPR TO QUELL STAPH INFECTIONS

If Zika and West Nile seem like exotic lurking risks, then staphylococcus is their more straightforward cousin. Staph just kills. With enough staph bacteria blooming in the body, the toxins they release can cause toxic shock syndrome. Basically, this means that the bacteria's by-products poison the body. Most staph infections are easily treated with antibiotics. But according to the Centers for Disease Control, every year in the United States an estimated 94,400 people are infected by an especially dangerous strain called methicillin-resistant *Staphylococcus aureus*, or MRSA, which results in almost 19,000 deaths annually. Often these infections are acquired in hospitals. Patients come to be cured but instead get infections that kill them. Hospitals try to cut down on MRSA infections with strong sterilization procedures. Researchers hope to combat MRSA with CRISPR.

Writing in the journal *Scientific Reports*, scientists from the Max Planck Institute in Germany describe the results of their genome-wide CRISPR screen searching for the human factors that let staph kill. First, they narrowed their exploration to the one staph toxin that does the most damage, the poison alpha hemolysin (the name literally means "blood kill"), and to the specific kind of human cells that die in droves during a staph infection, the myeloid cells that give rise to many types of white blood cells of the immune system. Would knocking out any single gene let myeloid cells survive staph's alpha hemolysin?

Of course, one target is the cell receptor on myeloid cells that alpha hemolysin specifically attaches to. The ADAM10 receptor has already been the focus of drug development against staph, and this study reconfirmed that when this gene is turned off, the staph toxin is less effective in killing myeloid cells. But the study found a handful of other genes the toxin needs, all of which are involved in presenting ADAM10 on the cell surface. In other words, the toxin needs to land on ADAM10, but it also needs these genes that help to position its landing pad.

We've been talking about targets, but if you remember, there is another side to this targeted-therapy story line. Once we know the target, the second step is figuring out how to keep the target from doing its dirty work. With CRISPR, it may someday be possible to simply turn off or replace these target genes in the human body, but as you've read, this strategy still faces a number of technical, regulatory, and safety hurdles. Until then, doctors have to come at the problem sideways, using any number of creative strategies to affect these genes or their expression. One of the promising strategies that drug designers hope to use to tweak gene expression doesn't go after genes at all, at least not directly, but influences a chemical in the body that decides how a gene manufactures a protein.

The information in your genes is carried out of the cell's nucleus by messenger RNA (mRNA) to other places in the cell where the DNA blueprint is actually manufactured into a protein. That process breaks down when the mRNA gets clogged up by something called microRNA. Think of microRNA as a layer of spackle on top of words written in Braille—the mRNA's information is covered up and can't be read.

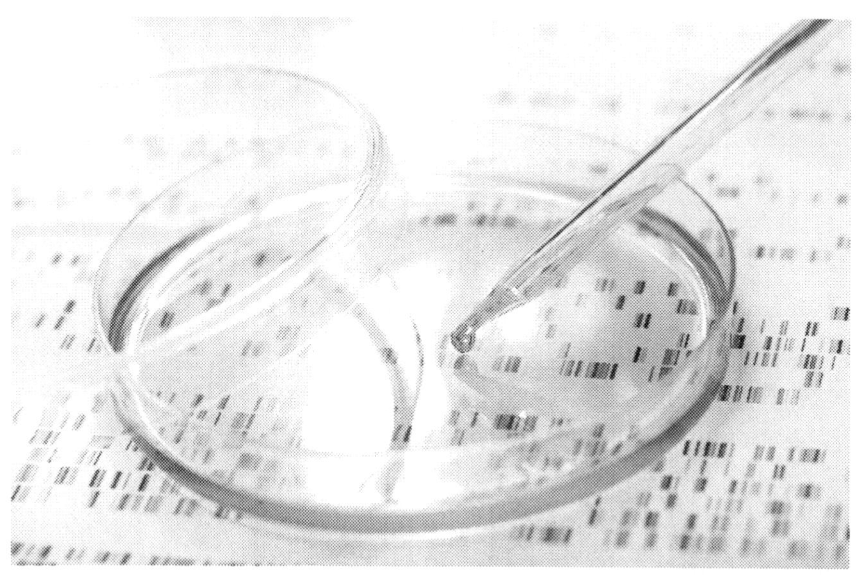

How would this complicated process treat disease? Well, by synthesizing a microRNA specifically designed to pair with a certain gene's mRNA, doctors might be able to tamp down the manufacture of disease-causing proteins that come from defective genes. Conversely, with *less* microRNA, they might be able to turn up a gene's activity, which is even more difficult to do with existing strategies. (For example, it would be lovely to amplify the activity of the gene p53, which is a known cancer suppressor.) One strategy is the use of molecular "sponges" that soak up microRNA before it finds its mRNA target. Another approach uses CRISPR.

The February 2016 issue of *Scientific Reports* describes how it works. Like anything in the body, microRNA must be manufactured. And, as with everything in the body, the blueprints for microRNA live in our genes. The *Scientific Reports* paper shows how Chinese scientists used CRISPR to knock out the genes that manufacture microRNA. Of course, they first did this in vitro, but then they also succeeded in mice. When they introduced a CRISPR-Cas9 system to mice, it nixed the production of the targeted microRNAs long into the animals' future. When microRNA was impaired, production of its protein targets went up. Making microRNAs into drug targets has become an entire field of medical research. This study shows one way to do it, and it implies that not only could CRISPR someday be used to effectively turn off target human genes, but, in a nifty double negative, it could also be used to turn off genes that turn off target genes.

However, even with a target in one hand and a possible drug in another, there is no guarantee that the drug will work in the complex landscape of the body. For example, the aforementioned Chinese study cut down the amount of microRNA by 96 percent. But is a 96 percent reduction enough to make inroads against a disease? It's hard to know until thousands of hours and large amounts of money have been spent to develop a candidate drug.

A 2015 paper in the journal *Cancer Research* describes an approach that uses CRISPR to solve this problem. CRISPR isn't used only to cut out and replace genes; it doesn't have to *remove* a gene at all. It can simply cut into the genome, insert a new sequence, and then stitch everything back together, a bit like adding links to a chain. Researchers from the Swiss drug company Novartis used CRISPR to add a little extra DNA to a gene so that the proteins made from the gene carried tiny tags. These tags, called degrons, control the rate at which proteins decay. With this system in place, genes that make disease-causing compounds would continue making them, but the proteins would now be programmed to quickly self-destruct. However, if the researchers also added an antidote—a so-called shield compound that protects these proteins—they could dial up or down the precise amount of the potentially dangerous protein in the body. That's like exchanging a regular on-off light switch for a dimmer switch. Instead of using CRISPR to turn off certain genes, we can now turn down genes by degree. What percent of gene inhibition is needed to combat a condition—5 percent, 50 percent, 95 percent? In experiments on cancer cell lines, this system successfully matched the known effects of established therapies like PI3Ka and EZH2 inhibitors, implying that this new technique could have the same effect as existing drugs.

If you've made it this far, congratulations. This chapter is filled with enough new science to stun a computational biologist. But the takeaway is simple. In addition to being used as a drug, gene-editing technologies like CRISPR are being used to help us *discover* new drugs. By knocking out thousands of genes, CRISPR can show us which genes a disease needs to survive or cause damage. Then, by tweaking the genetics of disease cells or human cells, CRISPR can show us how to design drugs that turn up or down the manufacture of these genes. Importantly, drug design using CRISPR doesn't actually depend on using CRISPR to edit genes in the human body, a therapy that still faces many challenges—it can be done in a petri dish in the lab. This gene editing is here now and is leading to treatments that will save lives.

CHAPTER TWELVE

SHOULD WE ALTER DNA?

The idea that you would affect evolution is a very profound thing.

—JENNIFER DOUDNA, PHD

In late 2014 Junjiu Huang, PhD, a gene-function researcher at Sun Yat-sen University in Guangzhou, China, led a team that injected a CRISPR-Cas9 system into human embryos. This gene-editing program was meant to correct genetics that cause the dangerous blood disorder beta thalassemia. The change, if it worked and if the embryos could develop into viable babies, would result in heritable changes in the babies' genomes, meaning that the researchers would have introduced artificial genes that would become part of the circulating universal human genome. Our genome is the product of millions of years of evolution. The work of Huang and his fellow researchers had the potential to unbind us from evolution overnight.

Fortunately or unfortunately, depending on your perspective, the experiment failed. Not only was the experiment intentionally performed on nonviable embryos, but of the 86 embryos tested, only 28 were successfully modified, and only a few of those contained the correct replacement gene. Writing in the journal *Protein & Cell* (several leading journals including *Nature* and *Science* had turned down publication of the study on ethical grounds), the researchers reported that the corrected gene worked in only 4 of the 86 embryos. To attempt the same with viable embryos would require nearly 100 percent success. The group also found many off-target modifications, meaning that CRISPR had found its target sequence elsewhere in the genome and so had made the same genetic replacement in genes that had nothing to do with beta thalassemia.

This study, which was meant to pave the way for future applications of CRISPR in human patients, instead became a cautionary tale. The technology wasn't ready for prime time. Nor was much of the world ready for this technology.

As you've read, there are three ways CRISPR or other gene editing technologies might be used with human cells. One is to edit adult cells that have been removed from the body, making them into a kind of medicine before reinfusing them into patients. Another is to edit cells *inside* the body, correcting or silencing malfunctions in adult tissues. The third is to modify embryonic or germ line cells—the sperm and egg cells that pass on the genetic blueprint for every other cell in the developing body.

This third use is especially attractive, as it would let doctors correct horrific genetic diseases before a baby is even born *and* ensure that the baby could not someday pass on these genes that create diseases to future generations. Gone would be cystic fibrosis, sickle cell anemia, Huntington's disease, Duchenne muscular dystrophy, predispositions to certain kinds of cancer, and thousands of other heritable defects. Millions of people around the world could live longer, healthier lives. But for all its potential for good, the use of gene-editing systems to alter DNA in heritable cells is the most ethically challenging.

Just as CRISPR might correct color blindness, so too could it be used to engineer blue-eyed babies. Down this path lie what many have called "designer babies." And the question of where to draw the line on the spectrum from preventing disease and defects to editing genes for traits such as height, IQ, or athleticism is only the tip of the iceberg. The Center for Genetics and Society in Berkeley, California, is an educational nonprofit working to encourage responsible uses and effective societal governance of human genetic and reproductive technologies. That organization has posed seven reasons why the United States should prohibit human germ line modification with CRISPR, paraphrased below:

1. **PROFOUND HEALTH RISKS TO FUTURE CHILDREN.** Using CRISPR in germ line cells results in heritable changes, meaning that all progeny of the altered organism inherit these changes. It is impossible to predict the effects of these changes across the life span.
2. **THIN MEDICAL JUSTIFICATION.** Some of the conditions that are proposed as targets for germ line modification can be avoided by other means, for example by embryo-screening techniques. CRISPR and other methods of gene editing

may not be the only or even the best way to combat these conditions.

3. **TREATING HUMAN BEINGS LIKE ENGINEERED PRODUCTS.** It's not unusual for parents to try to shape their children's lives by encouraging them to take music lessons or participate in sports, but should they also have the ability or the right to shape their children via genetic engineering?
4. **VIOLATING THE COMMON HERITAGE OF HUMANITY.** As many differences as we have as humans, we all share the constraints of human DNA. The Center for Genetics and Society asks, "What happens if traits viewed as socially undesirable are merely problems to be solved in a system that makes 'fitting the mold' a biological possibility?"
5. **UNDERMINING THE WIDESPREAD POLICY AGREEMENTS AMONG DOZENS OF DEMOCRATIC NATIONS.** More than 40 countries including the United States have prohibited genetic alternations to germ line cells. However, our privately funded biomedical sector is independent of this agreement. Should US biomedical companies undermine the policies of countries?
6. **ERODING PUBLIC TRUST IN RESPONSIBLE SCIENCE.** Scientists and humanitarian workers already experience skepticism and resistance to important programs such as vaccine delivery. Experiments that could change the genetic inheritance of our species has the potential to further erode trust in science and medicine.
7. **REINFORCING INEQUALITY, DISCRIMINATION, AND CONFLICT IN THE WORLD.** In the organization's own words, "The determination of 'bad' genes that need to be replaced and 'good' genes to be introduced would reflect criteria set by the economically and socially privileged." This reality could allow genetic manipulation to become another form of discrimination.

These opinions are largely echoed by the research community. For example, in a commentary by five leading genome scientists published in the March 2015 issue of *Nature*, titled "Don't Edit the Human Germ Line," the authors wrote, "In our view, genome editing in human embryos using current technologies could have unpredictable effects on future generations. This makes it dangerous and ethically unacceptable. Such research could be exploited for non-therapeutic modifications. We are concerned that a public outcry about such an ethical breach could hinder a promising area of therapeutic development, namely making genetic changes that cannot be inherited."

Also, there are environmental concerns about the ability to make heritable changes in plants and animals. While these genetic alterations could wipe out disease-carrying mosquitoes, for example, the long-term effects on the environment are not known—and may not be known for years. This "try and see" approach to manipulating ecosystems has a long history of failure. For example, some species of Asian carp that were introduced to the United States for use in cleaning commercial aquaculture ponds have escaped into river systems and adjoining lakes, where they have gobbled up ecosystem energy sources, decimating native species (not to mention the farcical drawback of these jumping carp injuring boaters and water skiers and physically knocking a kayaker from a Missouri River race in 2010). Or take the introduction of European rabbits to Australia in 1859 by British landowner Thomas Austin, which has been one of the most significant factors in species loss on the continent. Or consider the introduction of the mongoose to Hawaii, which was meant to control the rat population at sugarcane plantations. The mongoose has instead ravaged Hawaiian bird populations.

We would introduce edited genes into ecosystems with the same good intentions as those who have introduced nonnative species—to introduce beneficial traits such as resistance to disease, elimination of disease vectors, and greater food production. But as in the famous Ray Bradbury story "A Sound of Thunder," in which the accidental killing

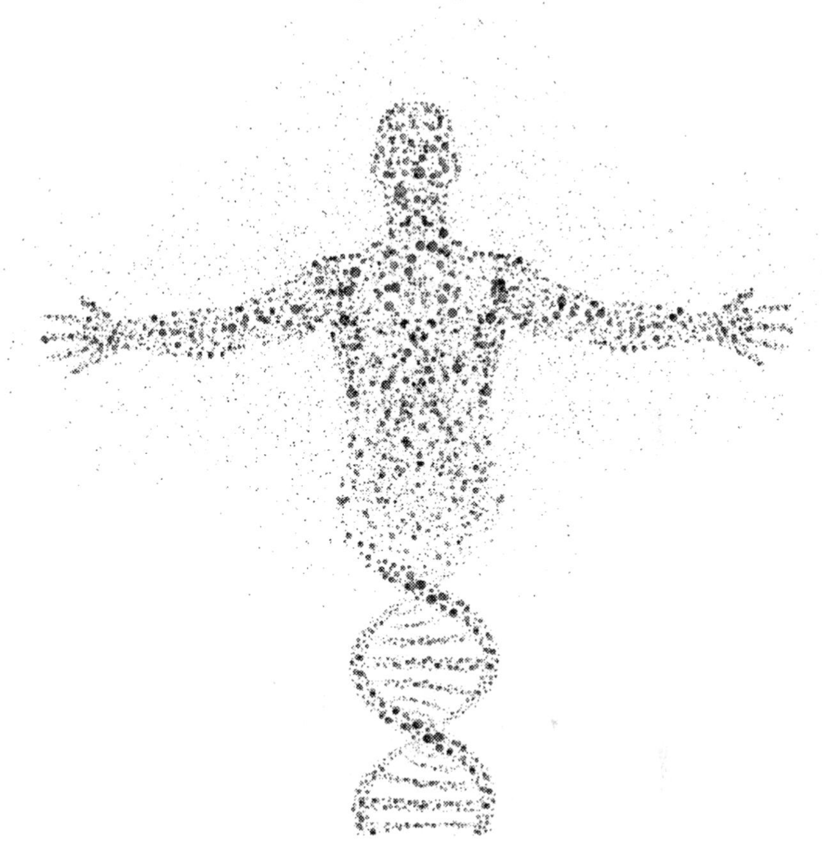

of a Cretaceous-period butterfly by a time traveler has dramatic unforeseen consequences for the present, it can be hard to predict the ripple effects of seemingly beneficial ecosystem changes.

Likewise, if genetic engineering can boost human health, it can also harm it. If history is a guide, it is only a matter of time before any new technology is used as a weapon. Albert Einstein, PhD, once said, "I know not with what weapons World War III will be fought, but World War IV will be fought with sticks and stones." Could weapons based on gene editing be the tool that returns society to the Stone Age, for example by famine or plague?

Einstein's concern is echoed by modern scientists. Bioethicist Gregory Stock, PhD, founding codirector of the Harris Center for

Precision Wellness at the Icahn School of Medicine at Mt. Sinai Hospital in New York City, says our concerns about human enhancement disregard the more likely evil that could accompany human gene editing: "New technology is probably already in the wrong hands. Create a super race? No, that is an extremely challenging and difficult thing to do," Stock says. "What I fear, however, are the people out there trying to do something really destructive and feasible with our new biotechnologies, like creating new plagues that will have the capability of killing many. There are people in the world bent on mayhem, and these are the people we really have to be wary of."

With anxiety building over the ethical and societal implications of CRISPR, the scientific community held an international summit in late 2015 to look at the various implications of human gene editing. More than 400 scientists, bioethicists, and members of watchdog groups from 20 countries gathered in Washington, DC, for the first International Summit on Human Gene Editing. Included were the three pioneers of CRISPR: Feng Zhang, PhD, of Harvard and MIT's Broad Institute, Emmanuelle Charpentier, PhD, of Max Planck Institute for Infection Biology, and Jennifer Doudna, PhD, from the University of California, Berkeley. Among other questions, the assembled experts all tasked themselves with deciding whether to propose a moratorium on heritable genetic modification. It was time to decide the near future of CRISPR in humans, go or no-go.

Here are highlights from the statement the summit produced:

- "If, in the process of research, early human embryos or germline cells undergo gene editing, the modified cells should not be used to establish a pregnancy."
- "Many promising and valuable clinical applications of gene editing are directed at altering genetic sequences only in somatic cells—that is, cells whose genomes are not transmitted to the next generation [essentially sperm and eggs] . . . Because proposed clinical uses are intended to

affect only the individual who receives them, they can be appropriately and rigorously evaluated within existing and evolving regulatory frameworks for gene therapy."
- "It would be irresponsible to proceed with any clinical use of germline editing unless and until: the relevant safety and efficacy issues have been resolved [and] there is broad societal consensus about the appropriateness of the proposed application . . . At present, these criteria have not been met for any proposed clinical use."

The summit committee stopped short of recommending a ban on germ line research, but discouraged its use until safety and efficacy issues have been resolved. In fact, many saw the summit as a late attempt to put a genie back into a bottle that had been opened with the completion of the Human Genome Project in 2003. When using CRISPR to snip genes from DNA, it's necessary to know the base-pair sequence of those genes, and that is what the Human Genome Project provided. And it has also led to another ethically challenging initiative aimed at manipulating human DNA.

HUMAN GENOME PROJECT-WRITE

Now, in the second decade after the human genome was first sequenced—which is to say, completely read—dozens of American scientists are banding together to *write* a new human genome from scratch. The project, called Human Genome Project–Write (HGP-Write), was announced in May 2016 in a letter published in the prestigious journal *Science*. The twenty-five scientists and business people who coauthored the paper proposed raising $100 million to use in synthesizing within the next decade a human genome of up to one billion base pairs (remember that a human genome has three billion) that could function inside living cells.

The project, spearheaded by George M. Church, PhD, professor of genetics at Harvard Medical School, plans to insert this artificial genome into a living human cell to replace its natural DNA in the hope that the cell will begin operating on instructions provided by the artificial DNA. The casing would be a human cell, but its actions and development would be artificial. In the future, HGP-Write would also create genomes of animals and plants. As outlined in the paper, the benefits could be many, including the ability to engineer cells that are resistant to viruses, the creation of therapeutic stem cells made with supercharged tumor-suppressor genes, new humanlike platforms for drug development, and the ability to modify species such as pigs so that they could be used to grow donor organs for transplantation into humans.

However, the ethical concerns may be as compelling as the potential benefits. In effect, the paper suggests building artificial life. Is it a good idea to chemically manufacture the human genome? If it proves possible to synthesize a genome of one billion base pairs, why stop there? Why not engineer the full three billion base pairs, which could be used to grow a parentless super child? Would project designers own this manufactured genome, and any viable cells or even humans that resulted from it? Most courts have ruled that a naturally occurring gene cannot be patented, but this would be a totally synthetic genetic creation—or would it? If it is simply a copy of what the original Human Genome Project had already determined and sequenced, would these genes be naturally occurring or synthetic? Could we engineer synthetic humans specifically to resist the rigors of space travel? If we can synthesize the genome, we could theoretically remake babies in the image of any existing DNA sample, say Albert Einstein or Gandhi or a departed loved one.

Again, the imaginations of science fiction (and comedy) writers have been only slightly ahead of the science: Will the ability to make a human from scratch lead to something like Stephen King's *Pet Sematary* in which beloved animals return from beyond, only slightly changed?

Or Woody Allen's film *Sleeper*, in which a futuristic society tries to clone its deceased leader from his only remaining body part, his nose? Or the film *Creator*, in which a scientist played by Peter O'Toole tries to clone his deceased wife? Promising research underway by top scientists is now bringing these fantasies closer to reality, minus, perhaps, the horror and the laughs.

DESIGNER HUMANS

Throughout human history, human ego and perhaps hubris have made it possible to see ourselves as distinct from nature, the only species gifted with consciousness and self-awareness. Unlike other animals, we don't possess super strength, superior running speed, claws, horns, or other adaptive physical features. We survive by our intellect. We are unique and special. And now, thanks to biotechnology, neuroscience, nanotechnology, robotics, artificial intelligence, and other fields of science, human intellect gives us the opportunity to separate ourselves from the fabric of nature in real and significant ways. Until now, we've tended to see ourselves as having risen above all other species—"how noble in reason, how infinite in faculty"—in short, more distinct from other forms of life than we really are. Now we have the possibility of making that perceived distinction literally so in many ways. It would be arrogant to believe that we could predict the consequences.

The promise is immense. CRISPR pioneer Dr. Jennifer Doudna has said that we "may be nearing the beginning of the end of genetic diseases." This can't be overstated: CRISPR and other gene-editing technologies could cure many of the maladies that have tortured humans since the rise of our species. But it also has the potential to end what is, strictly speaking, the human species. Are we not defined by the integrity of our genome? Without this integrity—with new, artificial genes coursing across generations in the wild type of humankind—would we continue to be "human" at all? The question now is not whether we could unbind ourselves from the genomes of our ancestors, but whether we *should.*

PART FOUR
HUMAN 2.0

There are no such things as incurables. There are only things for which man has not found a cure.

—BERNARD BARUCH

CHAPTER THIRTEEN

CAN WE PREVENT OR EVEN REVERSE ILLNESS AND AGING?

What a drag it is getting old.

—MICK JAGGER AND KEITH RICHARDS

Can human beings live forever?

It's not likely, but thanks to vaccines, antibiotics, improved sanitation, and access to safe and adequate supplies of food and water, life spans have lengthened significantly over the course of human history. The average life expectancy during the Cro-Magnon era was eighteen years. In the 1850s the average life expectancy for a European man or woman was forty-three years. Now, according to a recent report from the United Nations, a woman's life expectancy in the United States is eighty-one years, seventy-six for men. Those

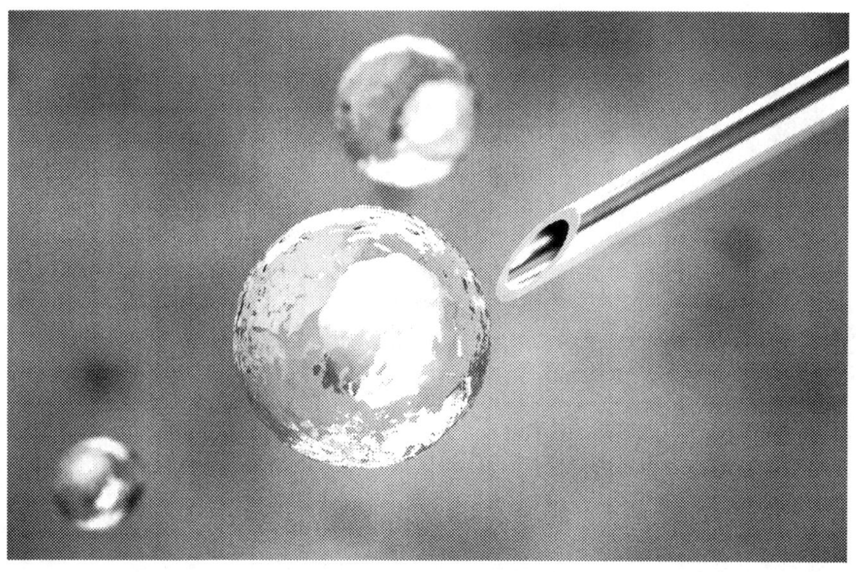

numbers are up from seventy-three and sixty-six only fifty years ago. That's a striking contrast; we've added nearly a decade of life in just a couple of generations.

We are now poised to take the next great leap in extending life span. Although aging is one of the most mysterious processes in biology, scientists are starting to unravel its secrets. These secrets, as with so many mysteries of the human body, lie in cells. Mutating cells and damaged molecules define aging. Cell-based therapies are now being developed to combat this damage, remanufacturing the human body into a machine our ancestors would barely recognize. Call it Human 2.0.

As a natural phenomenon, exceptionally long, healthy life is not unheard of. Accounts of "super-agers" have existed for years.

Jeanne Louise Calment was born in Arles, France, in 1875—the same year that Leo Tolstoy was publishing *Anna Karenina* and a year before Alexander Graham Bell patented the telephone. When she was 13, Calment met Vincent van Gogh, who famously lived and painted in Arles for a time, when the artist was buying canvas in her uncle's shop.

Calment never expected to live an extraordinarily long life, and she didn't intentionally do anything to foster her longevity. She did marry a wealthy businessman and never had to work, but she was very active, playing tennis, swimming, hunting, roller-skating, and riding her bicycle. Over the years, she buried her daughter, who died from pneumonia in 1934 at age 36; her husband died of food poisoning in 1942; and her only grandson was killed in a car accident at age 36.

Calment was born a quarter century before humans could fly, but she lived to see a man walk on the moon and the construction of the International Space Station. She lived through World War I, the Nazi occupation of France during World War II, the Korean and Vietnam wars, and the global AIDS epidemic. In 1975, the year she turned 100, Calment finally gave up bicycling. Instead, she spent time walking around Arles, going from house to house to thank those who had sent birthday wishes for her centennial.

When Calment was 110 years old and finally slowing down physically, she moved into a local nursing home. She stopped smoking when she was 117 but continued eating two pounds of chocolate weekly and drinking port wine right up until her death in 1997. She was 122 years old.

Jeanne Louise Calment isn't the only super-ager. A minute portion of humans live far beyond the normal life span and remain vital well into their tenth decade and beyond. At the Longevity Genes Project at the Institute for Aging Research at Albert Einstein College of Medicine in New York City, Nir Barzilai, MD, is studying a group of these exceptional centenarians. What do these super-agers know about healthy living for a long life?

Not much, it turns out.

"We found something completely different," says Dr. Barzilai. "They don't eat anything special, and they don't exercise regularly. As a matter of fact, 50 percent of them were obese, 50 percent did not exercise at all, and 60 percent of the men and 30 percent of the women were smoking. One study subject died at the age of 110; she had been a smoker for 95 years."

Instead, they were probably protected by something in their genetic makeup. "We have found several genotypes that have functional relevance in our study population," Dr. Barzilai says. "These genes actually change something in the body that is associated with longevity. We found through sequencing that these people had the cholesterol ester transformer protein gene, or CETP, the so-called longevity or 'Methuselah' gene that raises HDL cholesterol, the so-called good cholesterol. CETP was overrepresented in those who were 100 years old and older."

But even super-agers eventually die, and when they do, it tends to be from the same diseases as non-centenarians.

"For me, this suggests that it's the delay in aging, not protection from one or more specific diseases that is the critical factor," Barzilai says.

If you are not lucky enough to have a super-ager's CETP genes, you'll be glad to know that Barzilai is looking for and designing drugs to mimic their effects, targeting the cellular degradation, persistent low-grade inflammation, and DNA damage that underlie most age-related chronic disease. By reducing or reversing these cellular markers of old age, Barzilai hopes to literally slow down the aging process and thus the onset of age-related diseases.

One of these wonder drugs may already exist. The drug is metformin, probably the most widely used diabetes drug in the world, which was first synthesized in the 1920s from compounds derived from French lilac plants. The drug was first approved for type 2 diabetes in Canada in 1972; the US FDA followed suit twenty-two years later. Metformin is a well-known, well-studied, inexpensive drug, costing just pennies per pill. Dr. Barzilai will soon be testing its ability to forestall the onset of several age-related diseases and extend the healthy life span.

Specifically, over the course of the next few years, this trial, Targeting Aging with Metformin (TAME), will test the drug at fourteen aging centers around the United States. Dr. Barzilai, who is conducting the trial with the assistance of the American Federation for Aging Research, expects to enroll 3,000 volunteers between seventy and eighty years of

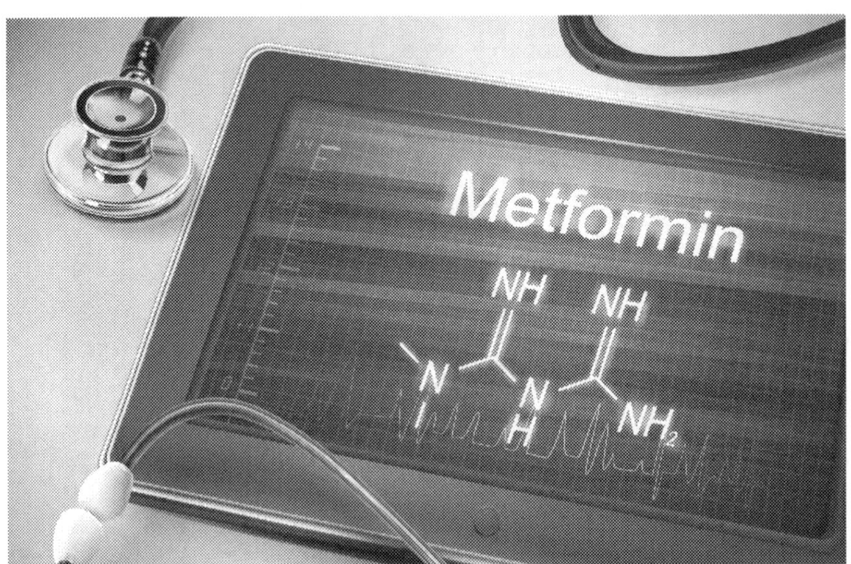

The molecular formula for metformin, a widely used diabetes drug that is being tested as a possibly effective anti-aging medication.

age and follow them for seven years. Half would take metformin daily, the others a placebo. The objective, of course, is to see if the development of diseases other than diabetes is delayed by metformin, as it was in the animal studies that led to FDA approval for this human trial.

Metformin works by reducing glucose production in the liver and increasing the body's sensitivity to insulin. This has obvious relevance to diabetes, and because it has been widely used, doctors and scientists have had the opportunity to make a number of interesting observations.

"People on metformin get 30 percent fewer cases of cancer—and this applies to almost every kind except maybe prostate cancer," says Dr. Barzilai. "There is also an indication that metformin prevents cognitive decline." Even more remarkable is a 2014 report by British researchers in the journal *Diabetes, Obesity and Metabolism* that a retrospective analysis of 78,000 people with type 2 diabetes found that those who took metformin lived 15 percent longer than people *without diabetes*. Think about that. People *with* diabetes, taking metformin, lived longer than people *without* diabetes. Will people *without* diabetes benefit from the drug as well? Dr. Barzilai expects to have an answer in a few years.

TELOMERES AND AGING

Another antiaging target is the diminishing telomeres at the ends of human chromosomes. Every time a human cell divides, it has to precisely replicate the three billion base pairs of its DNA. Any given cell in the body will divide fifty to seventy times during a person's lifetime. Because replication starts at the ends of the long strands of DNA, all this copying is hard on a chromosome's tip and tail, and so chromosomes have protective "caps" at each end, called telomeres. They're made up of long repeating lengths of nucleotide bases (the components

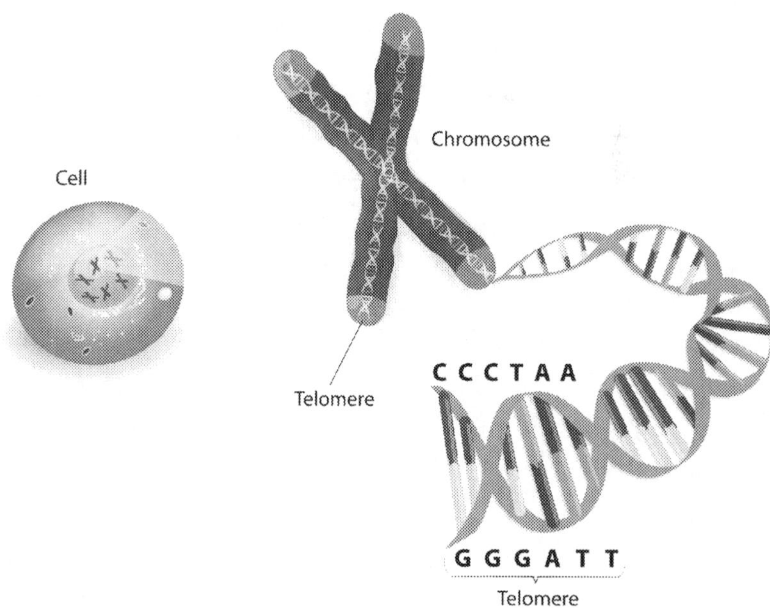

Chromosomes are found in the nucleus of the cell. Each end of a chromosome is covered by a protective cap of DNA with long repetitive sequences of nucleotides of CCCTAA paired with GGGATT. Telomeres shorten with cell division, eventually leading to cell senescence and death.

of DNA) whose sole function is to protect the rest of the DNA strand from damage, allowing cells to divide without losing genes off the ends of chromosomes. But telomeres have their own limits.

When we're born, our telomeres measure about 8,000 to 10,0000 nucleotides long, but they get shorter with each cell division. And as telomeres shorten, the ability of the cell to repair itself and maintain its normal function diminishes. Eventually, telomeres are degraded to a point at which the cell can no longer divide. Scientists call this the Hayflick limit. Cells expend telomere length in order to replicate until they reach the Hayflick limit, at which point they either become senescent or they die. It's this process of telomere shortening that's thought to be the biochemical basis of cellular aging.

Many researchers now think that telomere length, or lack thereof, is a molecular measure of aging, and that finding ways to lengthen and protect telomeres can turn back the biological clock in these cells, and thereby extend human life. In fact, shorter telomeres, which are more common in older cells, have become associated with a broad range of aging-related diseases, including stroke, dementia, cardiovascular disease, obesity, osteoporosis, and diabetes.

The good news is that there's an enzyme called telomerase that is able to repair and help lengthen telomeres, allowing cells to reproduce indefinitely. Even better news is that lifestyle changes may increase the activity of telomerase and help lengthen your telomeres—a pair of small studies reported that men who were identified as low risk for developing prostate cancer had an increase in their telomerase activity and telomere length after undertaking changes in diet, exercise, stress management, and social support.

"These new findings indicate that telomeres may lengthen to the degree that people change how they live," says lead study author Dean Ornish, MD. "Research indicates that longer telomeres are associated with fewer illnesses and longer life." Further, the scientists learned that the more the men changed their behavior by adhering to the recommended lifestyle program, the more dramatic their improvements in

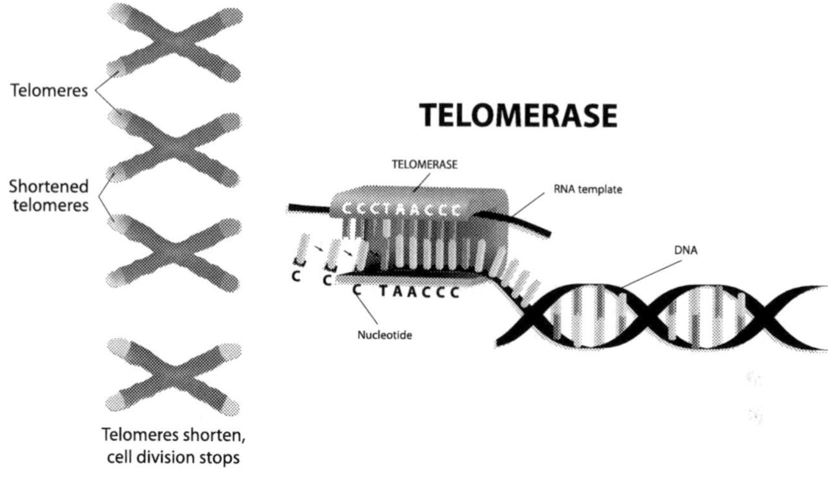

Telomerase is an enzyme that repairs the ends of chromosomes (telomeres) that are damaged with cell division and aging.

telomere length. These studies suggest that we can actually refashion our fate at the basic DNA level by making the right lifestyle choices.

But telomerase and telomere length have a dark side; cells that reproduce indefinitely are cancerous. One feature of cancer cells is their immortality. Where the slow shortening of telomeres puts the ticking clock of the Hayflick limit on a cell's life span, almost all cancers undermine this system, usually by restarting production of the enzyme telomerase. With more telomerase, cancer cells are able to rebuild their telomeres' TTAGGG sequences more quickly than they break down, allowing them to blow right past the Hayflick limit and continue dividing. (Drugs targeting telomerase are an exciting area of cancer research.)

As a result, we have a problem similar to that of cell senescence. The process of telomere decay is an essential guard against cancer but also a cellular driver of aging. It's theoretically possible to prevent some of the consequences of aging by somehow boosting telomerase activity, which would, in turn, lengthen telomeres, but there's a potential Catch-22: the attempt to make normal cells "immortal" could endow cells already tipping toward cancer with the same ability. Using telomerase therapy to live longer could literally be the death of you.

So how might we prevent the aging associated with shortened telomeres without jumpstarting cancer? Researchers recently reported the ability to lengthen telomeres in the lab. Cells with just a 10 percent increase in telomere length divided many more times in the culture dish than did untreated cells—about twenty-eight more times for skin cells and about three more times for muscle cells.

That approach is being pioneered at Stanford University. In 2015, Helen M. Blau, PhD, professor of microbiology and immunology and director of the university's Baxter Laboratory for Stem Cell Biology, "tapped the gas pedal" of cells with shortened telomeres. According to the university, Blau's goal was to offer a one-time telomere boost while continuing to allow all cells to continue coasting to a stop. Her approach targeted the enzyme telomerase, but in a unique way. Messenger RNA (mRNA) carries the sequence of a gene from DNA in the nucleus to where the gene's information is used to manufacture a protein in other parts of the cell. Blau's procedure used sophisticated genetic engineering to manufacture mRNA in the sequence of the gene called TERT, which codes for telomerase. With artificial TERT mRNA flooding the manufacturing machinery of the skin cells in which Blau tested the technique, the cells bloomed with telomerase, which the cells used to lengthen telomeres with new TTAGGG sequences. Importantly, the engineered mRNA was used and then quickly degraded, fully dissipating 48 hours after the boost.

"Now we have found a way to lengthen human telomeres by as much as 1,000 nucleotides, turning back the internal clock in these cells by the equivalent of many years of human life," says Blau. Results published in the January 2015 issue of the *FASEB Journal* showed that cells with telomeres lengthened by the procedure were able to divide 40 more times than untreated cells. (The Hayflick limit for human somatic cells is in the range of fifty to seventy divisions.) Importantly, this method of telomere lengthening is not thought to push cells as much towards too much cell division—that is, cancer.

"This new approach paves the way toward preventing or treating diseases of aging," says Blau. "There are also highly debilitating genetic diseases associated with telomere shortening that could benefit from such a potential treatment." The researchers are now testing their new technique in other types of cells.

So while short telomeres may predispose us to certain degenerative diseases, long telomeres that allow cells to keep dividing may not always be a good thing; very long telomeres are associated with an increased risk for cancer. It would seem that telomeres need to be a Goldilocks length: not too long and not too short—just right. Trouble is, we don't know what "just right" is for telomere length, and it may be different for each of us.

EXTENDING HEALTH AND LIFE WITH REPLACEMENT ORGANS

These approaches targeting plasminogen activator inhibitor-1 (PAI-1), telomeres, and senescence use drugs to heal or clear cells. Another approach to extending the health span is simply to replace an aged or otherwise failing organ with a new one. Of course, this strategy isn't new.

As any middle school student learns, the heart is the body's main pump, and when it malfunctions, the body dies. The first serious attempts to extend life with new hearts began with transplantation. In early 1967 Richard Lower, MD, of the Medical College of Virginia transplanted the heart of a young person killed in a car crash into the chest of a baboon to see if the human heart could be removed and restarted in a foreign body. Lower's radical experiment worked, although the baboon later died when its immune system rejected the human tissue.

Later that year, Christiaan Barnard, MBChB, a South African internist who had performed one of that country's first kidney transplants and had learned cardiac surgery at the University of Minnesota,

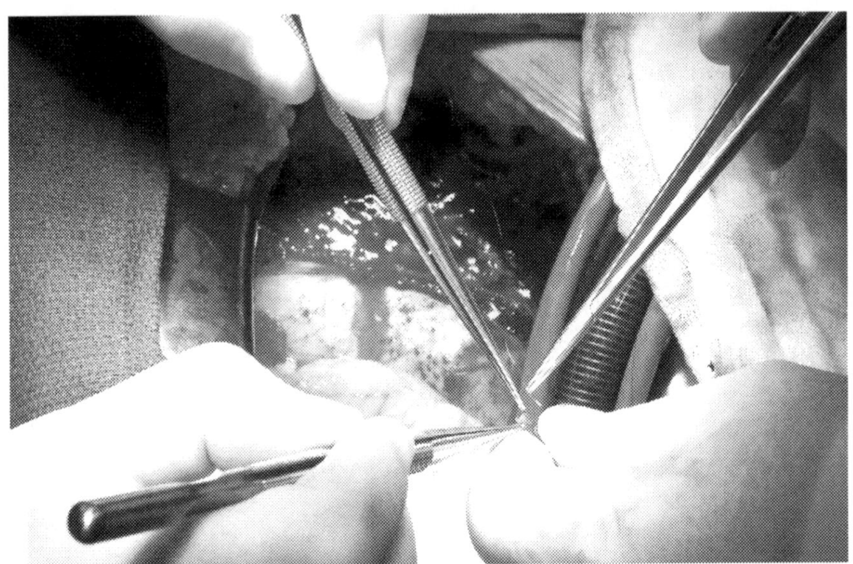

Doctors performing a human heart transplant.

did what no one had ever done before. He removed a diseased heart from a cardiac patient, Louis Washkansky, and replaced it with a heart harvested from a car-accident victim. Although Washkansky survived only eighteen days, eventually dying of pneumonia, he dramatically showed the world that a heart transplant was possible, paving the way for many surgical and pharmacological advances that would eventually make heart transplants a common therapy for end-stage heart disease patients. More than 100,000 heart transplants have been performed worldwide, with almost 90 percent of the heart recipients surviving for a year and almost 80 percent surviving at least five years.

The problem we now have is that there are not enough donor hearts available. It's estimated that about 4,000 heart patients are now waiting for donor hearts in the United States. These people are close to death, some suffering from long-term heart failure, others from heart-muscle disease, irreversible heart injury from coronary artery disease, congenital heart disease, and multiple heart attacks—conditions that can't be treated by any other medical or surgical means. Unfortunately, according to the American Heart Association, there are only about

2,500 donor hearts available each year, which means that thousands of patients die waiting for a heart that never comes.

Doris Taylor, PhD, director of regenerative medicine research at Texas Heart Institute, believes that with stem cells it will eventually be possible to build a new fully functional biological heart that can be implanted in humans. This, she says, will allow people to effectively overcome the finality of heart disease, the leading killer of men and women worldwide, and extend life perhaps by decades.

Taylor's approach, using a process called whole-organ decellularization, has succeeded in growing functioning rat hearts. The two-day process begins with removing all the cells from a heart that will be used as a model, leaving only the extracellular matrix, the framework between the cells, which is so devoid of genetic markers that it should not trigger rejection in a new body. This matrix serves as the scaffold for growing a new organ. After being cleaned with sodium trideceth sulfate, the same detergent used in baby shampoo, the matrix is infused with a mixture of progenitor stem cells from the hearts of newborn rats.

Four days after seeding, Taylor's team saw heart contractions. Eight days later, to the surprise of everyone, the hearts were pumping. The nickel-sized rat hearts beat on their own, just as if they were inside live animals.

Taylor and other scientists hope to employ the same technique to build new human organs by decellularizing hearts, livers, and other organs taken either from human cadavers or from larger animals such as pigs and coating them with the patient's own stem cells. Since these new organs are much less likely to be rejected by the body, there's no need for powerful antirejection drugs. This approach could lead to a virtually limitless supply of organs for transplantation that are every bit as functional as the ones that Mother Nature gave us.

BUYER BEWARE

Those who want to reverse the aging process should be mindful. From stem cell cocktails to telomere tests to untested and unregulated dietary supplements, snake oil salesmen and fly-by-night clinics have made big business out of the universal desire for antiaging therapies. We can understand the desperation from which they profit. Nothing is more certain than the march of time that engenders in many of us an existential dread of diminishment and decrepitude. Some of us are able to make peace with aging, but let's be honest: if there were a real way to slow down your body's internal clock, you'd jump at the chance. Who wouldn't? Now, with cell therapies backed by peer-reviewed science, that may be possible. Now as never before, we can look below the surface of the body's visible signs of aging to see the basic biology driving this process at the cellular and even molecular levels. When cells age, you age. Now new techniques are starting to deliver on the age-old dream of slowing, stopping, or even reversing this aging process.

CELL SENESCENCE

We've known for decades that cells divide for only so long before shutting themselves down and eventually committing programmed suicide, a process called apoptosis. Among other things, it's an essential defense against cancer. Once a cell endures enough damage, rather than continuing to replicate dangerously injured DNA, it becomes senescent—old and no longer able to divide. Some senescent cells undergo apoptosis, while others are swept away by the immune system. Over time, unfortunately, the immune system becomes less efficient at

As cells move through their life cycle, they become senescent and contribute to the aging process.

clearing senescent cells, and so they start to build up in the body's tissues, including the heart, kidneys, liver, and perhaps in the brain as well.

These senescent cells aren't necessarily dormant or inert. While they stop replicating, they continue to produce a mixture of chemicals including proinflammatory cytokines, for example, that damage adjacent cells and cause chronic inflammation, which is closely associated with most age-related diseases. Eventually, the by-products and waste products of senescent cells do more than contribute to stiff muscles and achy joints. Exposure to chronic inflammation can eventually kill you in a variety of ways, including by causing cancer.

To see just how dangerous senescent cells are, biologists Darren Baker, PhD, assistant professor at the Mayo Clinic, and Jan van Deursen, PhD, chair of biochemistry and molecular biology at Mayo, used sophisticated techniques to see what would happen if they cleared mice of senescent cells.

"Cellular senescence is a biological mechanism that functions as an 'emergency brake' used by damaged cells to stop dividing," says Van

Deursen. "While halting division of these cells is important for cancer prevention, it has been theorized that once the emergency brake has been pulled, these cells are no longer necessary. Senescent cells that accumulate with aging are largely bad; they trigger inflammation and enzymes that damage organs and tissues, and therefore shorten the healthy phase of your life."

"We think these cells are bad when they accumulate. We remove them and see the consequences," says Baker. "That's how I try to explain it to my kids."

A 2016 paper in the journal *Nature* describes their results. Importantly, when they injected middle-aged mice with a compound called AP20187, one of a new class called senolytic drugs, the experimental drug searched out and removed more than half of the senescent cells from the treated mice. Over time, these mice remained healthier than the control mice that did not receive the drug. With AP20187 helping them to clear senescent cells, the mice had less fat accumulation, improved heart and kidney health, and fewer eyesight problems. They developed cancer less often and later in life and were more active. The treated mice, on average, lived eight months longer than those who did not receive the drugs, which in mice translates to a one-third longer life span. It's hard to read this list of effects without interpreting it as evidence that medicines can affect the pace of aging.

The team's experiment "gives you confidence that senescent cells are an important target," says Dominic Withers, PhD, a clinician-scientist who studies aging at Imperial College London and who cowrote an article for *Nature* that accompanies the Mayo Clinic report. "I think that there is every chance this will be a viable therapeutic option."

In fact, scientists are even further down this road of development and testing with other promising antiaging drugs. This includes repurposing drugs already used to treat other conditions. (If a drug has been approved for one use, the FDA allows doctors to use that drug "off-label" for *other* rational uses.) This is how the drug epoetin alfa (PROCRIT), which was developed for kidney disease, was later approved for use with anemia. Similarly, the drug finasteride, developed to treat enlarged prostates and originally marketed as PROSCAR, came to be used for male pattern baldness under the name PROPECIA.

The obvious advantage of repurposed drugs is that they have already been tested for safety in humans and so may jump to the front of the line, while a compound like the Mayo Clinic's AP20187 is just at the beginning of a long and costly process of safety and efficacy testing before it can be used. (The only restriction is that drug companies can't promote or market a drug for off-label use—a rule that has led to some massive fines against the likes of Pfizer and Eli Lilly, among many others.)

A 2015 study in the journal *Aging Cell* tested the repurposing of two drugs against senescent cells. Dasatinib (SPRYCEL) is a cancer drug used to silence the malfunctioning BCR-ABL fusion gene (on the Philadelphia chromosome discussed at length in chapter nine) in chronic myeloid leukemia. The BCR gene is also part of a chain reaction—a "signaling network"—involved in senescence. Dasatinib breaks a link in this chain, and the researchers showed that dasatinib effectively eliminated senescent fat cell progenitors. In this study, they combined dasatinib with the dietary supplement quercetin, a flavonoid found in many foods, including kidney beans, capers, dill, and cilantro. Quercetin is an antihistamine and anti-inflammatory, and the research showed that quercetin effectively killed senescent human endothelial cells (which line the human heart, blood vessels, and lymph system) as well as senescent bone marrow stem cells in mice.

When the researchers gave a combination of dasatinib and quercetin to elderly mice, cardiac function improved just five days after a single dose. A single dose of this combination in mice that had a limb damaged by radiation improved the function of the limb for as long as seven months. Younger mice given an ongoing regimen of dasatinib and quercetin were slower to develop age-related symptoms, including osteoporosis and spinal disc decay. The researchers noted in their study that the "results demonstrate the feasibility of selectively ablating senescent cells and the efficacy of senolytics for alleviating symptoms of frailty and extending health span."

In a follow-up paper published in the same journal in August 2016, the Mayo Clinic group explored the effects of an ongoing regimen of dasatinib and quercetin on heart disease. In an interesting twist, they compared aged mice to middle-aged mice that had especially high cholesterol. Both groups of mice had reduced heart function. But only the old mice, not those with high cholesterol, benefited from the combination of dasatinib and quercetin. If senescent cells are to blame, this makes sense: the drug combination was not meant to help *all* cardiovascular disease, so this study suggests that the cocktail was performing as expected, reversing heart damage by the specific method of clearing senescent cells associated with aging and not by some other happenstance benefit that might have affected high-cholesterol mice as well. This second paper, too, ends on an optimistic note: "This is the first study to demonstrate that chronic clearance of senescent cells improves established vascular phenotypes associated with aging."

Scientists are now trying to move from these mouse findings to testing in humans. Based on the Mayo Clinic work, a new company, Unity Biotechnology, was founded in 2016 in San Francisco to develop senolytic medicines to treat and eliminate age-related diseases and increase health span. Backed by several venture capital companies and a Chinese drug company, Unity will focus first on developing drugs that could prevent and maybe even cure arthritis, heart disease, or loss of eyesight.

TARGETING CELL SENESCENCE IN THE BRAIN

What good is keeping the body spry if the brain can't keep pace? The Alzheimer's Association estimates that more than 5 million people in the United States had Alzheimer's in 2016 and that by 2050 this number will rise to 13.8 million. Along with this increase in human misery will come a massive increase in the cost of caring for these people. By 2050 Alzheimer's will account for one of every three Medicare dollars spent. The likelihood of a person's developing this dementia doubles every five years beginning at age 65, reaching nearly 20 percent by age 85. Alzheimer's disease is the sixth leading cause of death in the United States and a leading cause of disability. It is the only disease in the top ten causes of death that cannot yet be prevented, cured, or slowed.

This dismal landscape also makes Alzheimer's and other forms of age-associated dementia especially attractive research opportunities—even a moderately successful drug could do an amazing amount of good. This is magnified by the fact that Alzheimer's tends to be a disease of the very old. If a treatment strategy could push back the onset of Alzheimer's by only a few years, it would lead to a dramatic increase in the number of older Americans who are able to pass from other causes, with their minds still intact. When doctors talk about "squaring the life span" or "extending the health span," this is what they mean. The goal is to enable us to live longer *in good health*, with a blessedly short decline just before we die.

We now know that Alzheimer's disease is not something that suddenly occurs in old age. Rather, it's a continuum of illness that starts decades earlier. With a cure for Alzheimer's remaining stubbornly elusive, many researchers are shifting their focus to preventing the disease with interventions that start in middle age. Accordingly, much of the current research is now focused on early brain changes and early diagnosis with the goal of developing treatments that will effectively halt the progression of this disease.

As you read in this book's chapter on cell therapies for neurodegenerative diseases, various uses of stem cells hold much promise against Alzheimer's. Recently, the use of senolytic drugs has become another attractive strategy against the disease. The question is: If cell therapies aimed at senescence keep the body healthier longer, could similar strategies work in the brain?

The first step to answering this question is ensuring that cell senescence, in fact, exists in the brain. The question is far from a no-brainer. That's due to the simple fact that neurons don't really divide. It's not that they are senescent; it's that you are born with almost your full contingent of neurons. Learning and other changes in your cognition are due to wiring these neurons into different netlike structures, not to the growth of new neurons themselves. Neurons, while they don't necessarily divide, are not senescent—and drugs that sweep inactive neurons from your brain would almost certainly be a very bad idea. But neurons are not the end of the story. Surprisingly, neurons are not even the most populous cells in your brain; that title goes to the glial cells that support neurons. And just as these glial cells can divide, so too can they senesce.

A 2015 review in the journal *Experimental Gerontology* points out that not only have researchers found the fingerprints of senescence in

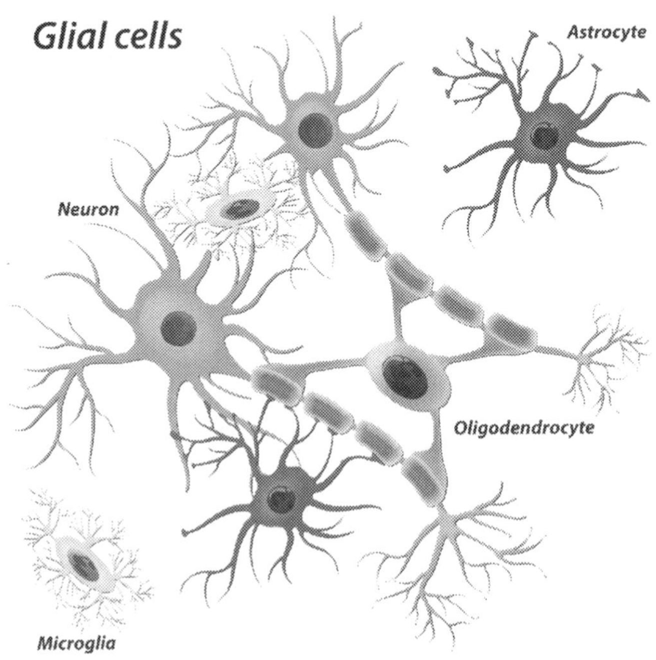

Glial cells are critical components of a normal functioning nervous system. They support, nourish, and insulate brain cells and nerve fibers.

the brain, including proinflammatory cytokines and growth factors, but also that increases in these chemicals over time "have the potential to cause or exacerbate age-related pathology . . . Senescent cells in the brain could, therefore, constitute novel therapeutic targets for treating age-related neuropathologies." In other words, not only is it perfectly possible for cells to senesce in the brain, but when they do, it engenders a world of age-associated conditions—and treating this senescence could help prevent or even reverse these age-associated conditions.

So far, so good. Science has defined an important problem, the age-associated accumulation of senescent glial cells in the brain. Unfortunately, this work is so recent that the studies starting to trickle upward from small journals tend to be limited to mechanisms of basic biology, rather than showing effects of possible therapies in mice or even in cells. Looking at the more established field of antiaging strategies in the body, it's pretty likely that antiaging drugs will be available for ailments of the heart well before the brain. But this does not discount the possible power of these cell-based antiaging strategies to prevent and even treat cognitive decline and even Alzheimer's. Now with a defined problem, a groundswell of science is gathering to solve it.

GENE EDITING TO STOP AGING

Many researchers are targeting genetics to treat aging. And as you've seen, many of the genetic inroads against aging flirt with a spectrum that includes cancer.

Take the gene MYC. This gene codes for a transcription factor, a substance that turns up and down the expression of a number of other genes. Basically, MYC is a regulatory switch. It's not very specific—researchers believe MYC may influence up to 15 percent of the body's genes, and its effects are varied and wide-ranging. However, if you group these genes by function, a couple of themes emerge, one of which is the cell-division cycle. Of course, anything that regulates the cell-division cycle is an attractive target for cancer. In this case, cancer hijacks MYC expression, speeding up cancer cells' process of replication. The gene is especially magnified in Burkitt lymphoma and may also be an important driver in a subset of cancers ranging from lung to cervix to breast and stomach. What this means is that, overall, turning up MYC expression is probably not a good idea.

But what happens when this gene is turned *down*? A team from Brown University decided to find out. It can be difficult to turn down a gene without turning it off completely, but one way to do it is to nix the gene on one chromosome and leave it alone on the other, creating a condition called haploinsufficiency. That's what the team at Brown did, damaging the Myc gene in mice and then breeding them with "normal Myc" mice. The offspring had one normal and one defective Myc gene—mice with half the expected Myc expression. The title of the team's article in the January 2015 issue of *Cell* sums up the results: "Reduced expression of Myc increases longevity and enhances health span." The mice's immune systems remained stronger for a longer time. The mice resisted the age-associated conditions of osteoporosis and cardiac fibrosis. "They also appear to be more active," the authors wrote. Also, the researchers saw that mice with low Myc burned energy more efficiently—both the easy energy of ATP (the basic unit of currency in cellular energy) and also in the metabolism of fat.

Again, MYC is a blunt instrument. Remember that while it mostly regulates things like cell cycle and energy, its umbrella extends over 15 percent of your genome, turning many things up and down. The mice in this study didn't seem to be harmed by this indiscriminate gene regulation, but it's hard to tell what the effects on humans might be.

That said, further studies are focusing on MYC, which is only one of a dozen or so promising antiaging targets. Other researchers are working with the TERT gene that makes the telomere-building enzyme, telomerase. And a group from Emory University has identified PGC-1a as a gene that helps mice resist aging-related diseases. Another group from the National Institutes of Health previously showed that reduction of the gene mTOR helped keep mouse tissues young and let mice live longer. And a Japanese group used the enzyme Klotho to affect the way mice used insulin, resulting in longer-lived mice.

This same strategy of identifying a genetic antiaging target and then turning it up and down in mice has now led to the first human trial of a promising compound called nicotinamide mononucleotide (NMN). While many antiaging strategies exist in the same cellular space as cancer, NMN shares its space with diabetes—and also with red wine. That's because NMN, perhaps like the resveratrol found in red wine and other sources, affects the production of a class of proteins called sirtuins. These regulate many processes associated with aging, including inflammation and how the body uses energy. When researchers at Queen Mary University of London gave NMN to mice, they found that it improved insulin secretion and islet function. That was in 2011. In 2016 a Japanese team recruited 10 volunteers to take the drug as part of a small safety trial. If the drug proves safe, the group hopes to expand the trial.

Other antiaging drugs have been down this road before. None of them have been proven to slow, stop, or reverse aging in humans. But now, with growing knowledge and new technologies, many leaders in aging research believe that the time has arrived—finally!—in which we'll realize the potential of gene-focused antiaging strategies.

ACCELERATING THE INNOVATION PROCESS WITH DOLLARS

Encouraged by these new approaches, investors and philanthropists are using money to speed up the process of antiaging research. Joon Yun, MD, president of Palo Alto Investors, a $1 billion private investment firm focused on health care, is spurring on the nascent life-extension field with their $1 million Palo Alto Longevity Prize, which actually offers two prizes of $500,000 each. The Longevity Demonstration Prize will be awarded to research groups that can figure out how to "hack the

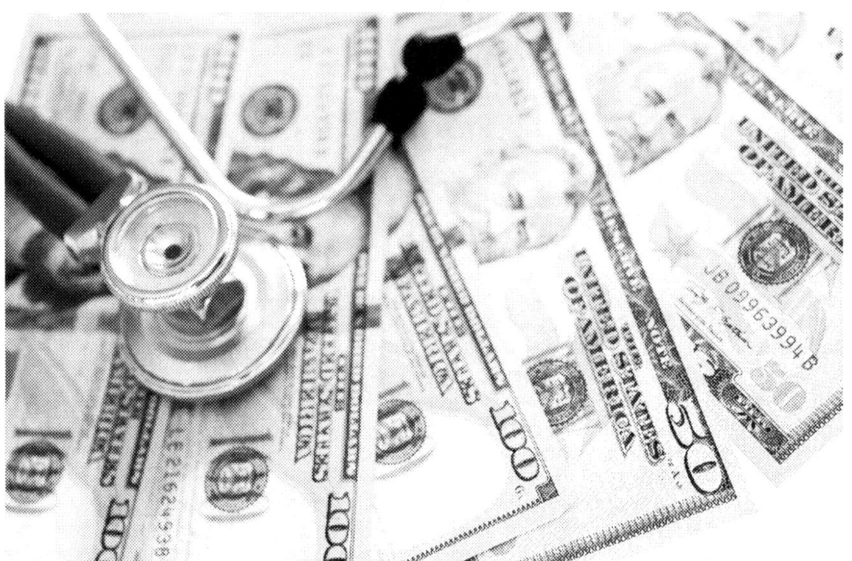

code" of aging in test animals. Winners must succeed in extending the life span of a mammal such as a mouse by 50 percent, the rough equivalent of an average American living to 120. The Homeostatic Capacity Prize will be given to the team that can turn back the age clock in an older mammal by restoring its heart-rate variability to that of a young adult. (The variations in the length of pauses between heartbeats are a common measure of a heart's "age.")

Yun, in an interview on Bloomberg TV, explained that therapies that keep the body in a youthful state, like those that he hopes will eventually come from the Palo Alto Longevity Prize, will ultimately reduce the costs of health care. "Think about your body as having a homeostasis system," Yun says. (Homeostasis is the ability to maintain a stable equilibrium.) "When we're young, it helps our body self-tune, and as we get older, especially beyond the age of forty, the system starts breaking down.

"With all the breakthroughs that we've had in science, there is no question that we can solve the longevity issue," Yun says. "We see the Palo Alto Prize as just a way to speed up the process. It will happen. It's just a race against time."

CHAPTER FOURTEEN

THE ROAD TO 100 PLUS

Fame comes and goes. Longevity is the thing to aim for.

—TONY BENNETT

In our efforts to improve human longevity, we harvested much of the low-hanging fruit during the twentieth century. Antibiotics, vaccines, and sanitation made inroads against disease. Improvements in food production and logistics helped many avoid conditions associated with malnourishment. The practice of giving birth in hospitals decreased deaths for both women and infants. And our understanding of the risks associated with smoking has cut down on deaths from tobacco use. As calculated in a 2002 article in *Science*, these advances have resulted in life span extension of about three months for every year from 1840 until 2010. *The World Factbook*, published by the CIA, estimates the life expectancy of a baby born in

the United States in 2010 at 78.24 years. (On the other hand, the life expectancy of a baby born in Mozambique in 2010 remained only 41.37 years.)

Despite developments like the obesity epidemic that thumb the scale against longevity, the US Census Bureau expects the trend of lengthening life spans to continue in developed countries, reaching the mid-eighties by 2050 before plateauing somewhere in the low nineties by the end of the century. That said, these numbers simply describe life *expectancy.* Even today some people live longer than others. And while the genetics you happen to inherit from your parents certainly influence life span, science is showing that lifestyle choices can shrink or expand the life span almost as much. An example of this emerged from a study of Seventh-day Adventists living in California. This religious denomination puts great emphasis on healthy living. Its members avoid tobacco, alcohol, and mind-altering drugs and maintain a vegetarian diet. A 2001 paper in the *Archives of Internal Medicine* reported a survey of 34,192 California Adventists that showed "diet, exercise, body mass index and previous smoking habits" coupled with the use of medicines when necessary "can account for differences of up to 10 years of life expectancy."

In other words, with the US Census Bureau estimating that the *average* life expectancy of babies born in the United States will reach into the nineties, people who make the right lifestyle choices should eventually expect to live past 100. This is the road we are on—the road to 100 plus. Of course, along this road, we must pass our nineties, and thanks to the ongoing work of researchers at University of California, Irvine, we already have a pretty clear picture of what it takes to live into the tenth decade of life.

THE 90-PLUS STUDY

Leisure World is a gated community of single-story row houses surrounded by manicured lawns built around a nine-hole golf course in Seal Beach, California. Today, over-fifty-five communities are commonplace, but Leisure World, built in 1960, was the first major planned retirement community in the United States. Today more than 9,000 people, almost all over age fifty-five, live there. Some of these residents are *way* over fifty-five.

Ruth Stahl is one of them. In her twenties, Stahl worked in a warplane factory where her job was to apply glow-in-the-dark paint to instrument dials in the cockpits. Ruth was sprinkled with so much radioactive paint that she glowed when she left work in the evening. Now, at ninety-six, she drives a green VW Beetle, walks three miles a day, and does shoulder stands in her yoga class. Her friend Jane is a couple of years older and smoked a pack of Pall Malls every day for nineteen years. For Irene, it was Chesterfields, and while she gave up smoking years ago, she still drinks scotch before dinner. Irene is 106.

How do they do it? It's partly blind luck. Every day we roll the dice against disease and injury, and some people continue through the decades without rolling "snake eyes." However, there are choices you can make that give you the advantage of playing with loaded dice, like exercise and a healthy diet—and some that are a lot more surprising. While it's easy to see the end results of these well-known age-extending activities in people like Ruth, Jane, and Irene, the real action takes place below the surface. Everything these women do—and that *you* could do—to live younger than your calendar age comes down to keeping your cells healthy.

In previous chapters, we have explored the idea that cells can be medicines—the right cells used in the right way can help the body cure itself of many conditions. But all these new and innovative uses of cells as medicines are attempts to fix things after they are broken—putting Humpty Dumpty together again, as it were. This chapter is about ensuring that he doesn't fall off the wall in the first place. From cancer to dementia to overall physical decline, supporting the health of your cells may be the best medicine of all.

We know this because Ruth, Jane, Irene, and thousands of other super-agers have volunteered to take part in a unique research project.

In 1981, researchers from the University of California, Irvine, mailed surveys to nearly 14,000 older adults living at Leisure World inviting them to participate in what has become known as the Leisure World Cohort Study. Over the years, as these residents aged and in many cases passed away, the researchers realized they had a unique opportunity to ask why others continued to thrive as they lived into their eighties and beyond. To date, more than 1,600 people have enrolled in an extension of the study known as the 90-Plus Study, and their participation in such great numbers has helped us reach some conclusions about what helps humans live longer.

Here are some of the key findings of this ongoing study:

GET YOUR HEART RATE UP. Adults in the 90-Plus Study who exercised for at least 45 minutes a day most days of the week were 27 percent less likely to die within an eight-year period than those who exercised less than 15 minutes daily. And yet, those who exercised as little as 15 minutes a day lived significantly longer than participants who were completely sedentary.

There is also a strong crossover between the physical health promoted by exercise and brain health. A 2012 study published in *Archives of Neurology* tested the physical abilities of 629 adults, average age 94, of which a quarter had been diagnosed with dementia. The question, as in the 90-Plus study, was whether

there were measurable differences between participants with and without dementia. To find out, researchers tested the volunteers' grip strength, standing balance, ability to quickly walk 13 feet, and ability to complete five chair stands (standing up from a chair and then sitting down). It turned out there were indeed dramatic differences between participants with and without dementia—and it certainly wasn't all in their brains. The strongest association was with walking speed—those unable to walk quickly were 30 times more likely to have dementia. Subjects with even minimal slowing of their walking speed (less than or equal to 1.5 seconds over the distance) were four times more likely to have dementia.

This is not to say with any certainty that decreased physical activity *causes* dementia—perhaps it was dementia that had caused mobility to fall and not the other way around. Or perhaps decreased mobility and dementia were both symptoms of another underlying condition. But further work shows that regular physical activity, even at an advanced age, improves blood flow to the heart and brain, so that brain cells receive more nutrients. There is also strong evidence that exercise—and this can include something as basic as a daily walk—helps promote the growth of new brain cells and the connections between the cells. When it comes to the health of your cells, in terms of overall life expectancy or "squaring of the life span" (to live as well as possible as long as possible), exercise may be one of the best medicines.

DON'T THINK THIN. With all the emphasis our culture places on linking ill health to body fat, you might think that being especially lean would help people live longer. You would be wrong. In the 90-Plus Study, older adults who exercised yet managed to stay slightly "overweight," as defined by the body mass index (BMI), lived longest. Specifically, participants with BMI in the 25 to 30 range (considered "overweight") lived longer than those with BMIs of 18.5 to 24.9, which is considered "normal." Of course, these "overweight" people also lived longer than people with BMI greater than 30, who are considered "obese" (and being obese at age 21 is associated with a *much* shorter life span). The reasons that slightly heavy people live longer aren't exactly clear. Perhaps weighing more is a marker of better overall nutrition. Maybe the longer lives of the overweight are due to the ability to resist age-associated wasting away of muscle and other tissue, which can lead to frailty. Whatever the cause, carrying a few extra pounds while continuing to exercise was associated with longer lives.

SOCIAL ENGAGEMENT MATTERS. No crossword puzzle or brain-training software can replace the benefits of being social. Getting out of the house and being around other people is associated with happiness and optimism, which are shown to have a protective effect on health. Moreover, navigating social situations forces the brain to work *hard*. For those of us used to socializing, it may not seem like a brain-busting chore, but unlike the compartmentalized skills that can be tested and perhaps trained by puzzles and other means, being social requires our brains' full involvement. For example, a 1997 study in *Nature* points out that the size of the human brain diverged from that of our nearest ancestors at the same time that we started to develop complex social structures. Whether a larger brain led to civilization or vice versa isn't clear, but, either way, the interplay between the brain and society is responsible for much of what makes us human. Processing faces, inferring the intentions of others, understanding irony and humor, anticipating reciprocal and competitive behaviors, and making judgments of right and wrong are among the higher-order skills that are special to the human brain. The 90-Plus Study shows that continuing to engage in stimulating social situations also maintains our brains, leading to better cognitive health *and* a decreased chance of mortality.

A GOOD DIET MATTERS. We already saw some of the results of diet in the longer lives of people who are mildly overweight. But it also turns out that what and how much you eat matters.

In the 90-Plus Study, people who ate smaller portion sizes lived longer and were less likely to be obese. It also helped to swap processed foods for fresh ones, especially plant-based foods such as vegetables, fruits, legumes, and whole grains. Some of the other findings pointed more to what people *didn't* consume—for example, older adults who ate more poultry and fish tended to eat less red meat. While the cause is unclear—that is, whether fish and poultry are helpful or red meat is harmful—these people lived longer.

CAFFEINE AND ALCOHOL. Speaking of what you choose to put in your body, the Million Woman Study published in 2009 in the *Journal of the National Cancer Institute* showed that even small amounts of alcohol consumption increase the risk of breast cancer. About one drink per day was associated with a 12 percent increase in risk. Clear connections between alcohol and cancer are also found in liver, esophageal, head and neck, and colorectal cancers. Still, the 90-Plus Study showed that older adults who drank moderate amounts of alcohol lived longer than their nondrinking peers. The reason may have to do with alcohol's cancer risk being overbalanced by its benefits—for example, people who drank small amounts of alcohol were more social and light consumption improved mood. As you have no doubt read elsewhere, there are also cardiovascular benefits associated with light alcohol consumption, possibly magnified if the alcohol consumed contains resveratrol, as does red wine. The 90-Plus Study found many of these same benefits with coffee—moderate consumption went hand in hand with a host of behavioral, social, cognitive, mood, and physical factors, and this data cluster also included longer lives.

YOUR CELLS AS ECOSYSTEM

Leisure World residents Ruth, Jane, and Irene are social. They exercise. Even though they are in their nineties or beyond, they are not pencil-thin. They eat well and drink moderately. And, of course, even though they smoked or were exposed to radiation earlier in their lives, they were lucky enough to have avoided the diseases associated with those experiences.

Living past 100 requires sidestepping the conditions that come with aging. The 90-Plus Study focused primarily on identifying the lifestyle factors that can help older adults avoid dementia. But mixed in with the data of the 90-Plus Study are numbers that speak to our overall chance of mortality. The same lifestyle choices that help to keep the brain sharp also kept people alive, and not only by avoiding dementia. Older adults who took care of their cellular health were more likely to avoid other life-limiting conditions as well, most importantly cancer.

The first reason that good cellular health guards against cancer is obvious. Things like smoking and excessive alcohol use increase the rate of DNA mutations in your cells, and hence the chance of getting cancer. In fact, for more than half a century, we have considered cancer to be the result of mutations. From that, it follows that the longer we live and the more careless we are with our behaviors, the more chance we have of accidentally picking up a DNA mutation that causes cancer.

Unfortunately, over these same fifty years, we have recognized flaws in this model. For example, it seems logical that large animals should be more prone to cancer than small ones because they have many more cells and therefore more chances for gene mutations. Yet blue whales seem to have about the same cancer risk over their lifetimes as mice. This is called Peto's paradox: cancer risk should be higher for species that have more cells, but it is not. Also, if cancer were simply due to picking up a random mutation, then we should see risk increase at a steady rate as we age. Yet it does not. Instead, cancer risk is somewhat

Our health and ultimately our life span depend on thinking of our cells and our bodies as complex ecosystems made up of many different human and bacterial cells, influenced by genetic predispositions as well as outside environmental factors.

higher in young children, then rests at a constant low rate through middle adulthood before rising sharply in older adults.

So, there must be other factors that contribute to cancer. It is becoming clear to scientists that cancer is caused not only by cells with mutations but also by the health of all the cells that surround these bad eggs.

Think of your cells as a nice green lawn. If conditions are controlled, grass forms a natural guard against weeds such as dandelions. When the ecosystem of your lawn is healthy, any dandelion that does happen to get a toehold is outcompeted by grass that is more perfectly optimized to the ecosystem.

The same is true in your body: healthy cells are optimized for a healthy tissue ecosystem. In fact, they are *so* optimized that almost any change to these cells makes them less fit. When a random mutation happens to cause a cancerous cell, this "dandelion" is just not the right fit for its surroundings and is quickly outcompeted. In a healthy tissue ecosystem, natural selection keeps cancer in check.

But dandelions are expert at exploiting disturbed earth. When the ecosystem is *not* healthy, all of a sudden grass is not the best suited to its surroundings. Now the tables are turned, and the dandelions that happen to get a toehold can outcompete the grass.

In your body, when the tissue landscape changes, cancer cells may suddenly outcompete healthy cells. Cancerous mutations happen all the time. However, when the ecosystem of your body's surrounding cells breaks down, cancerous cells are able to exploit these new disturbed conditions. In this model, it is your overall cellular health and not just random mutations that influence the risk of cancer.

This "natural selection" view of cancer solves both Peto's paradox and the nonlinear increase of cancer risk. Blue whales have no more cancer than mice because, despite the greater number of mutations, the whales' healthy tissues keep cancer in check just as well as mouse tissue. And it is in old age, when tissues start to break down, that cancer risk spikes.

This also means that anything we can do to promote the health of our cells decreases our cancer risk. The longer we can avoid "disturbed earth," the longer our healthy cells will remain optimized to their healthy surroundings, and the longer they will outcompete cancer cells.

This isn't just a harebrained hypothesis. A 2015 article in the *Proceedings of the National Academy of Sciences* links cancer with age-associated inflammation. First, the paper reinforces the idea that cancer-causing mutations in fact make cells less likely to survive in healthy tissue. However, these same mutations make cancer cells *more* likely to survive when surrounding tissue is gently inflamed, increasing the odds of a tumor developing. And inflammation is one of the hallmarks of aging tissue. "The key transition here should be from therapies and drugs targeting malignant cellular phenotypes to therapies and drugs targeting cell fitness," the paper asserts. In other words, our attempts to target cancer have been based on a model in which mutations cause cancer.

Kill the cells with such mutations, and you have killed cancer. But this new evolutionary model of cancer implies that we can control the disease in an entirely different way. Instead of killing cancer cells, we could enrich the health of all the cells that compete against cancer cells.

Inadvertently, this is what Rose, Jane, Irene, and others in the 90-Plus Study have done. Regular exercise, good diet, and other lifestyle choices keep their cells healthy, and in turn, these healthy cells keep cancer cells from becoming the ecosystem's dominant species, as it were. It turns out that from dementia to cancer to cardiovascular disease and more, the road to 100 plus is paved with healthy cells.

PART FIVE
BIG MONEY, BIG DATA, AND THE FUTURE OF MEDICINE

The world has been very careful to pick very few diseases for eradication, because it's very tough.

—BILL GATES

CHAPTER FIFTEEN

BIG DATA LEADS TO BETTER MEDICINE?

What gets measured, gets managed.

—PETER DRUCKER

Before the doubleheader that ended the 1941 baseball season, twenty-three-year-old Ted Williams, the "Splendid Splinter," was batting .39955 for the Boston Red Sox. Eight at-bats later, he had added six hits to push his batting average above the .400 mark, finishing the season at .406.

If you pay much attention to this kind of thing, you know that Williams was the last major league player to win a batting title with an average of over .400. On the other hand, in 2016 no less than 19 major league players hit .400 or better for the season. The trick is that to win an official batting title, a player now needs 502 plate appearances in a 162-game season. (In 1941, the season was 154 games.) In 2016

none of the .400 hitters had more than 25 at-bats, and most were relief pitchers who only batted once or twice.

In short, statistics mean little unless you have a whole lot of data. The larger the sample size, the smaller the margin of error. Start flipping a coin and you might turn up heads on your first five tosses. But flip the coin 500 times, and the results will be very close to the *real* odds of a coin flip—50 percent heads, 50 percent tails.

Of course, the same is true of medical statistics. If 2 patients are treated while participating in a clinical trial and one is cured, does that mean the medicine is 50 percent effective? If neither is cured, does that mean the drug failed? In both cases, no. You can't tell if it was the medicine or luck that was responsible for the results. You might need to test 200 patients to know if a treatment is effective or safe. This is especially true when the result might be an incremental improvement rather than an all-or-nothing cure.

This is one reason why data is so important. With clinical trials, there is strength in numbers. That's why we have the National Cancer Database (NCDB), which is a repository of more than *thirty-four million* records that show a patient's diagnosis, how the patient was treated, and the result. The power of the NCDB has helped doctors decide how to sequence, dose, and combine surgery, radiation, and chemotherapy against many cancers.

But new treatments go beyond this link between disease type, treatment, and outcomes. New treatments depend on drilling down into massive amounts of data to identify the genetic changes that are actually driving a patient's disease in order to match those changes with medicines designed to target those genes. New initiatives are springing up to provide this data, adding layers to the NCDB.

ORIEN AND OTHER DATA-SHARING COLLABORATIONS

When you look at the night sky, you see the twinkling pinpricks of a million stars. Maybe you can even recognize the bright stars Rigel and Betelgeuse along the celestial equator. But it takes the overlay of interpretation to see these stars as part of the constellation Orion. The Oncology Research Information Exchange Network, or ORIEN, is spelled a bit differently from the hunter constellation, but the goal is the same—to recognize patterns amid myriad pinpricks of data.

ORIEN is a novel partnership that pools data from clinical trials at nineteen academic medical centers. Among other things, it collects data that allows researchers to go beyond comparing medicines with outcomes to explore the genetic factors that may connect the two. The data comes from three sources: the clinical outcomes of how cancer patients fared on a variety of medicines (like NCDB data), gene profiles detailing the DNA that made up their tumors, and samples of the tumors themselves, which can yield information about which proteins these cancers express. As of November 2014, ORIEN has gathered more than 100,000 samples. Each sample includes genetic sequencing showing the three billion base pairs that make up the DNA of a tumor.

What does all this mean?

First, it means that doctors are learning from consortia like ORIEN how to use genetically targeted therapies the same way doctors learned from the NCDB how to use non-targeted therapies like chemotherapy. It also means that we've accumulated terabytes and terabytes of data. Hidden inside this data are cures based on molecules, genes, chromosomes, and cells, and new initiatives are rushing to capitalize on this tidal wave of collected information. President Barack Obama's 2016 Cancer Moonshot initiative, currently led by former Vice President Joe Biden, is driving the aggregation of data from industry and academic medical centers. The 2015 Precision Medicine Initiative is asking patients for their data directly. The 2013 BRAIN Initiative (Brain Research through Advancing Innovative Neurotechnologies) is using big data to map the connections between neurons in the central nervous system. And the international Human Cell Atlas initiative, launched in 2016, is using genetic data to categorize and map the hundreds of types and subtypes of cells in the human body.

We've seen the straightforward way in which more data leads to new medicines: The more patients enrolled in clinical trials and the more we know about the genetic factors that influence a drug's action and a patient's outcome, the more confident we can be of the usefulness of new medicines. But clinical trials are only the start. There are many stories that show the importance of big data in biomedical research. Let's take a look at a few.

DRUG REPOSITIONING

Non-small-cell lung cancer (NSCLC) has become a perfect example of the benefits of personalized medicine. Therapies have recently been approved targeting gene mutations known as ALK and EGFR, and clinical trials are testing drugs against a handful of other common NSCLC gene rearrangements, including NTRK, ROS1, and others that can cause the condition. Unfortunately, progress against its cousin, small-cell lung cancer, hasn't kept pace. In fact, there is not a single targeted therapy approved for use in small-cell lung cancer, and it has been thirty years since any new treatment has moved the needle on survival rates.

Today, only about 5 percent of patients diagnosed with small-cell lung cancer will be alive five years later. One factor driving the dismal evolution of small-cell lung cancer treatments is its rarity. Only about 10 to 15 percent of lung cancers are small-cell, arising from the neuroendocrine cells that release chemicals like serotonin in response to signals from the brain. Drug companies have been less motivated to spend the decade of work and $1 billion needed to develop and test a new therapy for a patient population that is too small to make the drug profitable.

That's a major reason why Atul Butte, MD, PhD, director of the Center for Pediatric Bioinformatics at Lucile Packard Children's Hospital at Stanford University, turned to "drug repositioning." This strategy, discussed more generally in chapter fourteen, is an ingenious way of looking for existing drugs that might be effective in treating conditions other than the ones they were designed and approved for. Instead of developing a new targeted therapy to fight small-cell lung cancer, Butte hoped he could find one already in use with another disease.

As with many stories of big data, this one starts with a database—actually four of them. These critical information repositories allowed Butte to compare the genetics of healthy lungs to those with small-cell lung cancer and also to evaluate the effects of many drugs from inside and outside the field of cancer therapy.

He used the first two databases to look for ways in which small-cell lung cancer was different from healthy lung cells. Any single cell is filled with too much genetic "noise"—too many random genetic changes that are irrelevant to disease or health—for researchers to be able to draw any conclusions. But when millions of cancer cells are compared to millions of healthy cells, patterns begin to emerge. The magnifying glass of big data let Dr. Butte see which genetic changes were most common in small-cell lung cancer. These differences tended to fall into

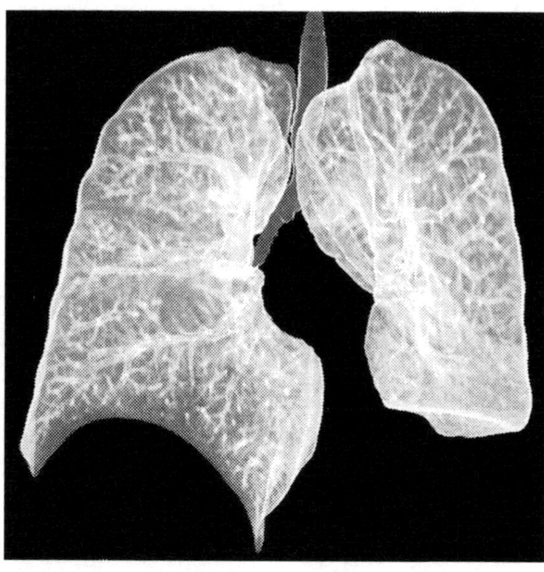

While lung cancer generally develops in one spot in a lung, it is lethal because it spreads to other parts of the body before causing many symptoms.

two categories: those involved in a cell's ability to sense and adapt to levels of calcium, and those involved in a neuroendocrine cell's ability to catch special signaling molecules that tell the cell what to do.

With that information, Butte turned to the drug database to ask which existing drugs are known to interfere with both of these systems. One such drug was imipramine (Tofranil), which had been around since the 1950s, when it was developed as a treatment for depression.

Imipramine is a small molecule that blocks the ability of neuroendocrine cells to receive signals from many kinds of hormones and neurotransmitters, including serotonin, norepinephrine, dopamine, acetylcholine, epinephrine, and histamine. If small-cell lung cancer inappropriately switches on neuroendocrine cells, it seemed logical that using imipramine to switch them off might be effective against the disease.

This deep data dive let Butte skip the time-consuming step of drug development. And because imipramine had already been FDA-approved for another use, there was no need for a costly phase I safety trial. Instead, Butte worked with Stanford oncologist Joel Neal, MD, PhD, to push the drug directly into a phase II clinical trial with small-cell lung cancer patients in California.

"We are cutting down the decade or more and the $1 billion it can typically take to translate a laboratory finding into a successful drug treatment to about one or two years and about $100,000," says Butte.

Repurposing drugs is not new. The hair-restoring drug ROGAINE was originally developed for hypertension. So was VIAGRA. Both modern uses were serendipitous discoveries—the drugs were repurposed when patients noticed these—ahem—side effects. Big data lets scientists like Butte replace luck with the intentional search for "side effects" that could possibly be viable treatments. And Stanford is certainly not the only place this computational drug repositioning is taking off.

Harvard researchers used a similar approach with the drug pentamidine, approved in 1937 to treat pneumonia, repositioning it to be used against metastatic kidney cancer. Researchers at Hunter College sifted through drug data and the genetics of disease to show that the type 2 diabetes drug metformin acted against cancers driven by changes in the genes EGFR and SGK1.

We take aspirin for headaches but also to lower the risk of heart attack and stroke. Big data is showing us that many drugs can be repositioned to treat conditions other than those for which they were designed. And just as this data lets us reposition old drugs, it is helping us pluck new ones from the haystack of drugs that have been developed but never used.

DRUG LIBRARIES

Today, many pharmaceutical companies are developing "libraries" of drugs that have yet to be matched with a purpose. One major category of drug libraries includes tyrosine kinase inhibitors (TKIs). TKIs keep proteins from getting the energy they need to become active. Many proteins can be switched off by specific TKIs. Drugs like imatinib (for chronic myeloid leukemia and other diseases; see chapter nine) and crizotinib (for non-small-cell lung cancer) are TKIs that turn off the proteins that cause disease.

What TKI will be the next life-saving drug? Well, for just over $4,000, the company Selleck Chemicals will send you a collection of 171 TKIs that you can test against whatever disease cells you please. BOC Sciences has a library of 118 completely different TKIs. Calbiochem offers 80, with free shipping for orders over $500! In fact, there are drugs developed to target more than 500 known tyrosine kinases, each turning off proteins that might have something to do with disease.

However, finding a match between one disease-causing protein and the TKI that turns it off isn't simple or easy. Unfortunately, genetic diseases are usually caused by a complex web of proteins. The challenge is to find the TKI that overlaps the greatest area of the web of disease-causing proteins while leaving proteins outside this web alone.

Here's where big data comes in. Until recently, the compound labeled AZD6244 was sitting on a dusty shelf in a TKI library developed by Array BioPharma in Boulder, Colorado. It was also included in a database called K-MAP ("kinase map") developed by researchers at the University of Colorado Cancer Center. At the time, K-MAP included about 2.5 million data points for 10,000 compounds. AZD6244 was just a drop in the ocean of untested TKIs. That is, until K-MAP pulled it to the surface. Researchers found a subset of colorectal cancers that seemed to be driven by the tyrosine kinase MAPK. (Yes, this is a disastrously confusing name: MAPK is the bad protein; K-MAP is the database tool.) The K-MAP database predicted that AZD6244 would play nicely with the immune-suppressing drug cyclosporine in inhibiting MAPK.

K-MAP pulled from the huge mass of data a streamlined hypothesis: AZD6244 plus cyclosporine would nix MAPK colorectal cancer. The combination went into testing, and the drug got a real name: selumetinib. After extremely promising lab results with cells and then with mice, the combination is now in human clinical trials.

There is a good chance that there are even more treatments waiting to be discovered within the 2.5 million data points of the K-MAP database. And this is one tool at one university. Across the country and around the world, big data is plucking drugs from libraries and pointing them at diseases.

DRUG DESIGN

In 1984 scientists at the National Institutes of Health conceived an ambitious project: to sequence the three billion base pairs of the human genome. The idea was funded in 1990, and the Human Genome Project soon included twenty universities in the United States, China, France, the United Kingdom, Germany, and Japan. The project sequenced the DNA base pairs of a handful of anonymous donors—primarily the genome of a man from Buffalo, New York—and then overlaid these strings of DNA to iron out most of the random genetic mutations. The project also tried to map the boundaries between genes, so as to

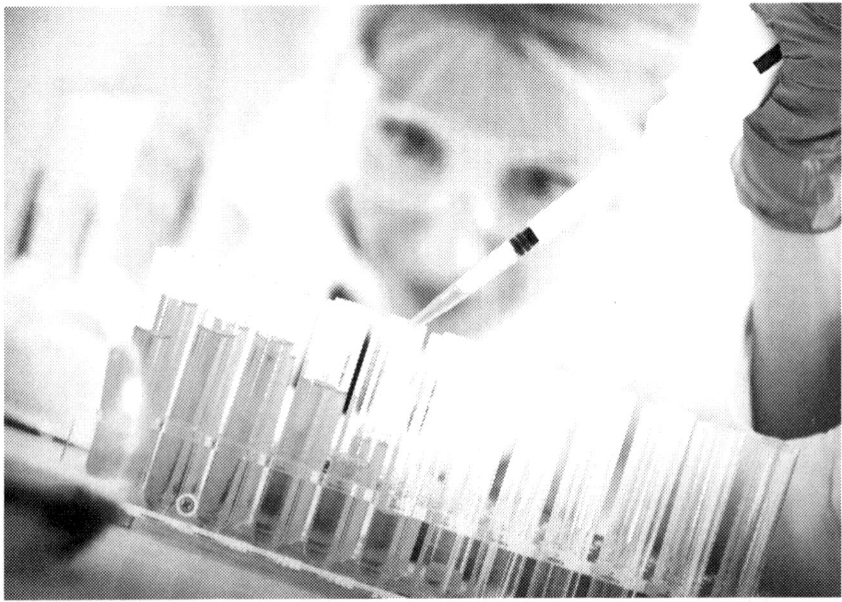

identify the sections within these three billion base pairs that form about 22,300 protein-coding genes.

It took ten years and $3 billion before researchers announced they had reached a draft of the "human reference genome." Today, in a stunning example of the power of new technologies, scientists and doctors can sequence the genomes of individual samples or patients in a few hours for about $1,000. But the human reference genome remains the roadmap. By laying a new sequence against what we consider normal, we can see important changes and differences. The results of the Human Genome Project, fueled by new technologies, are driving a paradigm shift in drug design. The trial-and-error strategy of blindly testing drugs against samples is obsolete. The new approach is to define and target the specific genetics that cause a disease.

The power of modern computing allows us to compare huge numbers of gene points in a short time. After using next-generation sequencing to define gene sequences, researchers compare these to "normal" sequences. A mismatch shows that a gene taken from diseased tissue is different from the same gene in healthy tissue, implying this gene may be influencing the disease and providing a possible drug target. To hit these targets, researchers pair next-generation sequencing with a technology called high-throughput screening. This system uses plates with thousands of tiny wells and automated systems that squirt test chemicals into these wells.

By this method, researchers can screen thousands of compounds against one kind of disease cell. Or they can screen one compound against many kinds of cells. Again, the result is data. In fact, there are nearly as many medical journal studies describing what to do with high-throughput screening data as there are papers describing findings that come from this technique.

Another drug-development technique based on big data seems even more futuristic. Remember the tyrosine kinase inhibitors you just read about that block proteins from getting the energy they need to become active? A problem is that TKIs only work with certain proteins, and there are many other proteins that researchers have pinpointed as likely causes of disease.

Take the protein known as RAS. In some cancers—including bladder, prostate, and likely many more—RAS-family proteins act as switches that activate out-of-control cell division, which is one of the defining features of cancer. And despite decades of effort, doctors haven't been able to do anything about it. There is no tyrosine kinase inhibitor, no gene therapy, no anti-RAS immunotherapy, and no other strategy to keep this known offender from going about its cancer-causing business. A 2014 article in the journal *Nature* noted, "Despite more than three decades of intensive effort, no effective pharmacological inhibitors of the RAS oncoproteins [cancer-causing proteins] have reached the clinic, prompting the widely held perception that RAS proteins are 'undruggable.'" Imagine how frustrating it is to know the cause of a disease and not be able to do anything about it!

Now there may be a way. When proteins transform from inactive to active, they actually change shape. And in some cases, it is possible to stick a monkey wrench in the gears of this process, causing it to grind to a halt. The monkey-wrench image is almost literal. Some proteins have a pocket that closes or twists upon activation, and specially shaped molecules may be able to attach to this pocket, holding it open and thus stopping the protein's activation.

What "specially shaped molecule" will fit a protein's pocket? That's a question for big data. The answer is "virtual drug docking." The right molecule will fit a protein like a key in a lock. But it may take sorting through literally millions of keys to find the right fit.

A 2014 study in the journal *Nature* describes the strategy that researchers from the University of Virginia, the University of Colorado, and Yale used to test these keys. They made a digital model of one of these RAS proteins and used sophisticated software to virtually bombard it with molecules. In all, they virtually docked 500,000 molecules, of which 88 seemed to have the right fit. They then took real samples of these 88 candidates to the lab, using high-throughput screening to test them against cancer cells.

From this screening emerged the drug BQU57. Actually, this is technically not a drug at all but an innocuous little molecule that just happened to have the right shape. But in mouse testing BQU57 started to look more like a medicine, slowing tumor growth in the test animals.

This study is a perfect example of the value of big data in drug development. Gene sequencing identified RAS-family proteins as drivers of a variety of cancers. Drug docking quickly sifted through 500,000 compounds to provide candidate molecules. And high-throughput screening tested 88 of these candidate molecules against cancer cells. The resulting data-driven discovery, BQU57, is now making its way toward clinical trials as the first treatment against RAS-driven cancers.

THE BIG DATA OF DISEASE PREVENTION

In this chapter, you've had a look inside the use of "big data" of DNA and drugs and clinical trials. But the most important application of big data might be outside the laboratory. Data that boosts your health may be no farther away than your pocket. The new cures cherry-picked by data mining are only one side of the coin. The other is the good old power of prevention.

Lifestyle influences the development of many diseases, and the data available via our smartphones is helping us humans track and adjust our lifestyles in ways that can counteract many dangerous conditions. What are your heart rate and blood pressure? How many steps have you taken today? What is your mood? What are your sleep patterns? Where are you in your fertility cycle? How many calories have you consumed today? How long did you exercise?

If you are using cloud-based apps to track these things, it may be disturbing to know that Big Brother is watching them, too. But along with this personal and sometimes passive data collection comes the ability to receive life-saving advice from a benevolent overseer of data.

Imagine knowing your precise risk for heart disease or diabetes. Now imagine working with your smartphone to make lifestyle changes that decrease this risk. If your risk is high enough, you may be able to treat conditions before they start. If your risk is low, you may be able to work with your doctor to avoid unnecessary medicines and procedures. Looking through the lens of data at the specific information of our own lives could help us prevent the very diseases that scientists on the other side of medical data are working to cure.

CHAPTER SIXTEEN

PHILANTHROPY DRIVES INNOVATION

Unless someone like you cares a whole awful lot,
nothing is going to get better. It's not.
—THEODOR SEUSS GEISEL, "DR. SEUSS"

Born in 1856, James Buchanan Brady was the second son of a New York saloon operator. He left school at the age of eleven and started working to support his family, eventually getting a job selling saws used for cutting railroad rails. He developed an eye for diamonds and other jewels, and as his success as a salesman grew, so did his diamond collection, earning him the nickname "Diamond Jim."

As renowned as he was for his business skills, Brady was also well known for his prodigious appetite. Historians note that his breakfast often consisted of a gallon of orange juice, a half-dozen eggs, pancakes, fish cakes, and chops, with food consumption increasing through the day to include dozens of oysters and clams, terrapin, lobsters, roasted meats, and a variety of game birds.

In 1912, already suffering from diabetes, kidney disease, and other ailments, Brady developed prostate difficulties that greatly affected his urination. Doctors in Boston and New York had already told Brady that because of his heart disease and diabetes, they could not help him with any surgical procedure on his prostate. So Brady headed to Baltimore for another opinion.

"A big, burly man with a huge head and a strong face appeared at my office," recalled Hugh Hampton Young, MD, a talented doctor at Johns Hopkins Hospital who had been appointed head of the urology department at the age of twenty-seven. "He wore a neat, well-fitting morning coat, but in his tie was a huge diamond, and diamonds also sparkled from his vest, watch chain, cuff links, and the head of his cane. He looked his nickname."

Dr. Young explained to Brady a new procedure he had developed that would help relieve pressure on Brady's prostate. Brady agreed to go forward, and on April 7, 1912, Dr. Young injected cocaine into Brady's urethra to numb it and then punched out three large pieces of tissue from the bladder neck. Shortly after his successful treatment, the ever-grateful Brady endowed what became the James Buchannan Brady Urological Institute at Johns Hopkins University.

"While his jewels, his motorcars, his famous dinners, and his intimate acquaintances among the great heads of railroads of the entire United States were sufficient to mark Brady as an extraordinary man, he was at heart simple and retiring, and one of the most considerate and generous men I have ever known," wrote Dr. Young.

Brady's gift was emblematic of the golden age of American philanthropy. At the time, John D. Rockefeller, Sr. was giving to education and religion, while Andrew Carnegie was building libraries. Even Henry Ford, who wrote, "The moment human helpfulness is systematized, organized, commercialized, and professionalized, the heart of it is extinguished, and it becomes a cold and clammy thing," donated at

Major American medical philanthropist, John D. Rockefeller, Sr. with his grandson David Rockefeller. John D. Rockefeller, Sr. supported medical education globally and David Rockefeller followed his family legacy believing that "philanthropy is involved with basic innovations that transform society." (Public image)

least a third of his fortune to causes including hospitals and museums. Around the country, philanthropists like Diamond Jim were using their wealth to build hospitals and staff them with what became a sea swell of talented researchers.

These donations turned out to be important investments for the nation, helping to push the United States past Europe to become the leader in biomedical research. Then the Great Depression caused a crash in philanthropic giving, and much of what remained was shifted to social services. However, the federal government took steps to fill the gap in medical research funding. In 1930, the Ransdell Act formalized the National Institute of Health (singular at that time), which was followed in 1937 by the National Cancer Institute and in the postwar period by several additional institutes, including the National Heart Institute and the National Institute of Mental Health.

Private philanthropy certainly didn't cease to exist in medicine, though. Howard Hughes, for example, started funding research in 1953. But a perception grew that the financial needs of medical research had been subsumed into a large federal system so that they were no longer a private-sector responsibility. Our taxes now funded government grants, and these grants funded research.

In 1980 the Bayh-Dole Act set the ground rules for an important shift in this system, allowing hospitals and universities to own their discoveries while reaching out to the private sector to develop and market their products. It was a beneficial partnership. Industry could effectively outsource basic research to academia, which was more than happy to continue focusing on deep science, with the added bonus that their discoveries could become part of the industry's discovery pipeline. This simple system held course through the early 2000s. Academia did the basic science, and industry brought it to market. Both benefited.

According to estimates by the Association of University Technology Managers, licensing activities in 2013 included 818 startup companies formed around patented medical inventions and other discoveries from American research centers. In that same year, royalties from medicines, devices, and techniques originally developed in academia totaled $23 billion.

However, this arrangement also meant that for the first time, taxpayer dollars had a purpose beyond igniting discovery. These monies could be used to make more money. And if recouping an investment was the ultimate goal, then public investment in research required a great leap of faith. That's because taxpayer dollars, distributed by the National Institutes of Health (NIH), paid for many projects that failed and for many more that led to seemingly inconsequential discoveries deep in the academic backwaters of biology.

In other words, the public wanted results for their investments, and those results weren't always easy to see. Basic scientific research often takes twenty years or so to pay off, if it pays off at all, and the

nation had what seemed to be more pressing concerns. Americans needed food and housing and access to existing medical technologies. Today, ten years of flat funding from the NIH have turned down the flow of speculative, basic science, not only reducing the current amount of lab work, but also thwarting the careers of a new generation of aspiring scientists.

We have entered an era in which only the most established researchers, proposing the surest projects, can hope to win the intense competition for grants. But if we're able to fund only very few studies, then doesn't that make sense? Shouldn't studies done by the best scientists have the best chance of leading to usable results?

The reason we should not follow this line of thinking is that less-common diseases, younger researchers, and high-risk–high-reward approaches are being left behind. Where as many as 35 percent of all NIH research grant applications were green-lighted a decade ago, only about 15 percent receive funding today. New technologies like next-generation sequencing, high-throughput screening, and cell-based therapies are driving exponential growth in the *potential* for biomedical discoveries that will change the world.

Dr. W. E. Bosarge with Cardinal Gianfranco Ravasi, head of the Vatican's Pontifical Council for Culture, after his appointment as Pontifical Council Admonitor and Senior Advisor for Regenerative Medicine and Adult Stem Cells in 2016.

W. E. Bosarge, PhD, is another energetic philanthropist who has an unwavering devotion to supporting regenerative medicine, personalized medicine, and cancer research and changing the regulatory paradigm. His contributions to medicine—as well as his many other charitable efforts, from supporting renovations and preservation of historical buildings and landmarks and constructing community improvements to honoring American veterans and promoting core American values—led to his being recognized around the globe by institutions such as the Vatican in 2011, with the Key Guardian Award, and again in 2013 and 2016, with the Key Philanthropy award. In 2017 he was recognized with the AACR Distinguished Public Service Award for helping to propel the American medical innovation engine.

THE RENAISSANCE OF AMERICAN MEDICAL PHILANTHROPY

The crippling of government-funded research is the bad news. The good news is that individuals and foundations are stepping in to fill the void. In some cases, this story of national redemption mirrors a story of personal redemption as well.

You know Michael Milken as the former Wall Street "junk bond king" who fell from grace (and went to prison for securities and tax violations) in the early 1990s. For Milken, it rained and then poured. In 1993, at the age of 46, he was diagnosed with prostate cancer and told he had twelve to eighteen months to live.

"A month before I was diagnosed, a good friend, Steve Ross, the CEO of Time Warner, had died from prostate cancer," Milken says. "Here was an individual with access to everything in the medical world, and yet he died from the disease. After Steve's death, I had to ask my doctor several times to give me a PSA test. He felt I was too young to get it."

Eventually, Milken talked his doctor into the test. When his PSA score came back at 24 ng/ml—anything above 4 ng/ml is considered worrisome—Milken and his doctor assumed there had been an error. Another PSA test followed by a prostate biopsy soon revealed that Milken had a Gleason score of 9—indicating aggressive, life-threatening cancer. And his lymph glands were 100 times normal size.

"I had been involved in the cancer research world for twenty years. I considered myself an extremely knowledgeable layperson," says Milken. "A first cousin on my mother's side died at thirty-one from a brain tumor. I had lost my father, stepfather, my mother-in-law, and my aunt to cancer. There wasn't a month that went by when a close friend or associate wasn't dying from cancer. I had plenty of people with cancer who interacted in my life. After my own diagnosis, I asked myself what was I going to do differently than my family and friends with cancer."

One thing Milken did was embrace holistic wellness along with Western medicine, trading meat for a healthy diet high in fruits and vegetables and meditating every day. For conventional medical treatment, Milken went to Johns Hopkins Hospital, where he was evaluated at the urologic center founded by Diamond Jim Brady. At the time, so few advances were being made in prostate cancer that promising researchers were steered away from the field. Milken decided to change that dismal state of affairs. After all, his life depended on it.

"I was confident I could bring order to the disorder that I saw," he said. "To me, this wasn't hard. I committed money for five years, with the major focus on recruiting top scientists to prostate research. I was focused on living and on solving the problem. We funded everything."

After doing his own medical due diligence, Milken decided to have hormonal treatment at Cedars-Sinai Medical Center in Los Angeles in order to starve his tumor of the testosterone that was fueling its growth. He followed up that six-pill-per-day regimen with two months of daily 3D conformal radiation therapy treatments, which were relatively new at the time. No cancer had spread from his prostate, and his PSA soon became undetectable.

While his disease had crested, Milken's philanthropy had just begun. To date, his Prostate Cancer Foundation (PCF) has made awards to over 2,000 researchers at more than 200 leading cancer institutions worldwide. These high-impact projects include clinical research to evaluate new drugs, novel treatment strategies, and basic research to better understand the biology of prostate cancer.

This financial support provided by Milken's group has led to many important advances, including the following: identification of the genetic changes that might cause cells in the prostate to become cancerous; interference with the development of blood vessels that deliver nutrients to cancer cells; identification of prostate cell surface markers that can be targeted to destroy cancerous cells; and development of analytical methods to identify the proteins in blood or the prostate that correlate with the aggressiveness of the cancer.

At the same time, Milken's PCF has funded key clinical trials in order to shorten the time between drug development and FDA approval. Of the nine new drugs for prostate cancer that have received FDA approval since 2002, six of them were supported by PCF funding at some point during their discovery or development. Each of these agents is poised to make a significant impact on the outcomes of patients with prostate cancer.

Today Milken's annual PCF Scientific Retreat at Lake Tahoe has become the most important international scientific conference on prostate cancer, attracting the most highly regarded scientific leaders in the field for three days of scientific presentations, poster sessions, and intense dialogue about clinical data, new discoveries, emerging treatment strategies, and policies to accelerate drug discovery and development.

Asked about the progress made on prostate cancer, Milken is decidedly upbeat. "I am quite optimistic," he says. "Due to the efforts of the Prostate Cancer Foundation, I am confident that we will eliminate prostate cancer as a cause of death long before we eliminate cancer in general as a cause of death. With prostate cancer conquered, we could then focus on another cancer."

OTHER SUCCESSES IN "BIG PHILANTHROPY"

Morgan was six years old when she was diagnosed with type 1 diabetes, a relentless autoimmune disease that, if not well controlled, can lead to blindness, amputations, heart disease, and kidney disease, among other complications. Like other parents confronted with such life-altering news, Morgan's father, David Panzirer, threw himself into learning everything he could about the condition. What he found was that the disease has no cure, requires constant glucose monitoring and meticulous insulin calculations, and can lead to death if not properly managed. Indeed, patients must make dosing decisions several times a day—without the benefit of a doctor—with a drug that could kill them if not administered properly.

Left: Morgan Panzirer was diagnosed with type 1 diabetes at age six.
Right: Morgan with her family.
(Photos courtesy of David Panzirer)

Panzirer quickly realized that the millions of people suffering from type 1 diabetes around the world needed better tools to ease the burden of managing this complicated disease.

Despite such a glaring need, Panzirer discovered that type 1 diabetes, with its almost two million American patients, wasn't nearly as attractive to the pharmaceutical industry as its cousin, type 2 diabetes, which affects at least twenty-five million people in the United States. This economic reality meant that while treatments for type 2 diabetes were in pipelines across the country, industry was much slower to focus on his daughter's disease.

Like all people with type 1 diabetes, Morgan soon became accustomed to needles pricking her fingers ten to twelve times a day to monitor her glucose levels and multiple daily injections of insulin to try to keep her blood sugar levels as close to normal as possible. Meanwhile, her father sprang into action. Shortly after Morgan's diagnosis, Panzirer was named a trustee of the Helmsley Charitable Trust, a private foundation created by his late grandmother, billionaire businesswoman Leona Helmsley. He knew that when the market doesn't provide the incentive to search for innovations, philanthropic groups have a unique opportunity to spur advances. And because research for type 1 diabetes treatments seemed to progress at a painfully slow pace, Panzirer and the team he helped assemble at the Helmsley Charitable Trust began to take a different approach—they wondered if researchers could learn to prevent the disease before it started.

Several new studies have been asking that question. Recently, a small clinical trial in Germany found that daily doses of oral insulin for children susceptible to type 1 diabetes can bring on vaccine-like immune responses, giving rise to the hope that such interventions could eventually prevent type 1 diabetes. Meanwhile, a large research initiative across Australia is observing pregnant women who have a first-degree relative with type 1 diabetes to identify environmental triggers that might influence the development of the disease early in life. And several multinational observational studies supported by the National Institutes of Health and others have been shedding light on how environmental factors such as diet and lifestyle might contribute to the onset of the disease.

"While these studies are encouraging," Panzirer says, "they are just the tip of the iceberg of what we have to learn." Existing efforts in prevention science have uncovered some critical pieces of the puzzle, but it will take a sustained commitment to this research and a collaborative effort to support future intervention studies in order to accelerate the path to prevention.

To date, the type 1 diabetes program at the Helmsley Charitable Trust has made over 400 grants totaling more than $350 million supporting a variety of efforts to improve the lives of people living with the disease. One new undertaking will help researchers in Europe explore safe and cost-effective ways to preserve the body's ability to produce insulin. Another avenue is a project to create a "bionic pancreas," developed by Boston University biomedical engineer Edward Damiano, PhD, whose teenage son was diagnosed with type 1 diabetes as an infant. Though not perfect, the bionic pancreas is one of several efforts that are a step in the right direction to begin to automate insulin delivery. Through critical financial help from Helmsley, the Boston researchers were able in 2015 to test their device with adolescents.

"Through several recent studies, including one reported in the *New England Journal of Medicine* in 2014, these systems have demonstrated strong potential to be a game-changer for people with type 1 diabetes," Panzirer says. "By providing better tools for patients and caregivers to manage this disease, we can help relieve some of the 24/7 burden of constant glucose monitoring and insulin injections, and actually improve health outcomes so those with type 1 diabetes can live longer, healthier lives."

Panzirer has had an amazing opportunity to have an impact on his own child's life by recognizing the power of philanthropy to accelerate advances for diseases that affect millions but might otherwise be ignored. "Through passion and dedication," Panzirer says, "the Helmsley Charitable Trust is investing in research that others cannot or will not fund. As Helmsley facilitates advances in device development, in the lab, and in clinics, it is taking crucial steps toward reducing the burden of type 1 diabetes and, we hope, one day preventing it altogether."

FROM HELPING HOMEBUILDERS TO HELPING KEEP PEOPLE ALIVE

Again and again, we see that committed philanthropists can create real inroads against disease. Take Bernie Marcus. When he was growing up in Newark, New Jersey, Marcus wanted to be a physician. Accepted into Harvard Medical School, he had to decline because he didn't have the tuition money. Instead, he went to nearby Rutgers, where he became a pharmacist.

Philanthropist Bernie Marcus, cofounder of Home Depot, has funded several biomedical research initiatives. (Courtesy of The Marcus Foundation)

Fast-forward twenty-five years: Marcus and a colleague opened a novel home-improvement store in Atlanta. Several years later, the creative duo opened dozens more stores and began growing into a national chain. Home Depot soon went public. In 2002 Marcus retired a billionaire and quickly switched his focus from drywall and hammers to medical philanthropy and his Marcus Foundation.

"Bernie Marcus likes to fund anything that is innovative and not likely to be supported by state support, NIH funding, or industry," says Fred Sanfilippo, MD, PhD, director of the Emory-Georgia Tech Healthcare Innovation Program and the medical director of the Marcus Foundation. "The bottom line is that Bernie wants to fund projects that really show clinical impact. That is his sweet spot. With his training as a pharmacist, helping people and providing health care has always been important to him."

Over the years, Marcus has given away more than $1 billion through his foundation. For example, in 2005, when his longtime friends Bob and Suzanne Wright, grandparents of a child with autism, founded the organization Autism Speaks, Marcus started the foundation off with a bang, donating $25 million. Since then, Autism Speaks has grown into the world's leading autism science and advocacy organization, funding research into the causes, prevention, treatments, and a cure for autism; increasing awareness of autism spectrum disorders; and advocating for the needs of individuals with autism and their families.

Recently, Marcus and his wife donated $25 million to create a neuroscience institute at Boca Raton Regional Hospital dedicated to the study and cure of Alzheimer's, Parkinson's, and multiple sclerosis. In May 2017, Marcus committed $38 million over five years to University of Colorado Anschutz Medical Campus to create the Marcus Institute for Brain Health. The institute will aim to improve programs for military veterans and their families, serving veterans suffering from traumatic brain injury, mental illness, and other ailments.

Ever the savvy businessman, Marcus is keenly interested in where his money goes and what it is expected to do. "Bernie can't understand donors who just write a check and walk away," says Dr. Sanfillipo. "He's developed an effective process for choosing grantees that is designed to create entities that will outlive him and his foundation's help. That's what a serious philanthropist does, and Bernie is very serious."

SEAN PARKER STEPS UP

Philanthropy certainly isn't limited to heavy industry and the established money of family foundations. Silicon Valley is in the game, too. For example, Sean Parker, who helped create the music-sharing company Napster at age nineteen and became president of Facebook at twenty-four, is now taking on the world of medical philanthropy with his estimated $2.6 billion fortune. In 2015 Parker established the San Francisco–based Parker Foundation with a $600 million grant to help support innovative medical research in the life sciences, along with public health and civic engagement.

Parker is not afraid to take risks. For more than a decade, he has been an active donor to institutions taking innovative approaches to cancer research and public health. For example, you read in this book about the ways in which cellular medicines are being used against severe allergies, and Parker has been a leading driver of this work. He recently pledged $24 million to create the Sean N. Parker Center for Allergy & Asthma Research at Stanford University and another $10 million to establish the Sean N. Parker Autoimmune Research Lab at UCSF. In the spring of 2016, Parker announced the formation of the Parker Institute for Cancer Immunotherapy with an investment of $250 million. These centers, in turn, are fulfilling a function that government funding never could—spurring the transition of cellular medicines based on the immune system from fringe studies to mainstream treatments.

Napster founder Sean Parker established the Parker Institute for Cancer Immunotherapy with a $250 million seed investment.

Henry Ford, the great American industrialist, once said that the best use of capital is not to make more money but to make money do more for the betterment of society. T. Denny Sanford, the munificent octogenarian from South Dakota who's widely considered one of the most generous philanthropists in the United States, has taken the early automaker's message to heart and then some. Throughout his decades of philanthropy, Denny Sanford has focused his gifting endeavors on health care, medical research, and education.

After amassing a multibillion-dollar fortune through his credit card business in South Dakota, Sanford directed a large portion of his philanthropic giving to a hospital located in the state. Now bearing his name, Sanford Health is one of the largest integrated health systems in the nation with 43 hospitals, nearly 250 clinics in nine states, and world clinic locations in Ghana, China, and Germany. Sanford's gifts total close to $1 billion and have allowed Sanford Health to launch several groundbreaking initiatives including Sanford Imagenetics. This first-of-its-kind model fully integrates genetic and genomic medicine

Philanthropist T. Denny Sanford. (Courtesy of Sanford Health)

into primary care like never before. This allows for more precision in prescriptions and in the screening and management of chronic disorders. His gifts also led to the creation of the Sanford Project, which is solely focused on finding a cure for type 1 diabetes. His generosity has also spawned programs and research into stem cell therapies, regenerative medicine, and cures for various cancers.

T. Denny Sanford understands that science and medical discovery is ultimately a numbers game. The more scientists—young and old—who are able to work because they have adequate funding, the better our chances of finding exciting and meaningful new medical discoveries.

BEYOND BILLIONAIRES

"More than funders, we can be partners building a bridge from academic research to industrial production," David Panzirer says. His words refer to one of two major friction points in the work flow of discovery. Even when academia makes a basic scientific discovery that has commercial promise, industry may be unwilling to invest in the early stages of translating the discovery into action. Too many promising drugs fail, and industry may be hesitant to take the risk. Philanthropy can help academic researchers take promising discoveries farther along the development pipeline—for example, into animal trials or even into phase I human trials. The more groundwork done by this academia-philanthropy partnership, the more likely that a pharmaceutical company can be tempted into committing the resources needed to push it across the finish line of FDA approval.

The second major friction point is in funding creative, innovative, sometimes long-shot ideas. In order to earn NIH funding, a project must already have shown promise, but it takes money for a project to get to the point where it *can* show promise. Seed grants and pilot funding from private philanthropy can ensure that outside-the-box studies

(often by young, unproven researchers) go forward in a way that could lead to larger NIH funding down the line. Because it tends to be flexible, risk-tolerant, fast-moving, and offered without the typical reams of red tape that comes with government monies, philanthropic funding is especially prized by scientists and continues to have a powerful impact on the field.

Philanthropic support doesn't have to come from billionaires. Across the country, thousands of donors and small foundations are searching out opportunities to drive the pace of research in diseases that would otherwise be left behind. These "pet projects" come from grateful patients, their families, and others affected by specific diseases. Maybe all it takes is $40,000 to test a new idea; increasingly, small grants from private philanthropies are the engine of innovation, while NIH grants follow behind to let researchers continue exploring the avenues that prove promising.

Today, you can even spend an amount of cash equivalent to your morning latte on the search for cures. Crowdfunding sites like Experiment.com, PetriDish, and angelMD let researchers post their proposals where microdonors can read, evaluate, and invest. These groups have helped fund such projects as an evaluation of nanoparticles against cancer, treating hearing loss in veterans, an imaging study of the brain on LSD, a study of the role of the aquatic fern azolla in climate change, and a technology-laden vest developed by famed neuroscientist David Eagleman, PhD, meant to allow humans to expand past the limitations of our five senses.

Yes, some of these crowdfunded projects are sensationalist, perhaps even a bit harebrained. And yes, many are doomed to fail. But some are also destined to succeed.

Researchers joke that in today's funding climate, they have to do the experiments first in order to earn an NIH grant. But on the other end of this spectrum are the tiny grants from organizations, individuals, and "crowds" that are betting a little on big results. While governments and the titans of industry continue to focus on big cures for big diseases, it may be the two-steps-forward, one-step-back lurch of spare-change philanthropy that makes the most difference in the years to come.

CHAPTER SEVENTEEN

CLINICAL TRIALS IN THE ERA OF CELLULAR MEDICINE

Mothers hold their children's hands for a short while, but their hearts forever.

—UNKNOWN

In March 2012 Julie Gibson's daughter, Jenna, was pale and just not feeling herself. Julie took Jenna to a pediatrician. Right away, the doctor knew it was more than a cold and ordered blood tests.

"As soon as the doctor got the blood results back—it was just two hours—she was calling us and telling us to go to the ER at Seattle Children's Hospital, immediately," says Julie. Jenna was diagnosed with leukemia. Within 24 hours, she was on chemotherapy. But unfortunately

for Jenna, chemotherapy wasn't enough. Jenna needed a bone marrow transplant.

"One reason we were so devastated when we found out she was going down the transplant road is that you need a matched donor, and the first place they look is siblings. Jenna happens to be adopted, so we knew a sibling donor match was not an option. And second, the percentage of donors that are non-Caucasian is tiny. Jenna's non-Caucasian," Julie says.

Julie was right to be worried. An extensive search produced no bone marrow donor who matched Jenna's blood profile. And there were no other FDA-approved therapies available.

But there was one last option. A clinical trial offered at the hospital was exploring another source of bone marrow, stem cells from donated umbilical cord blood. These cells are promising for transplantation because for the most part they haven't had time to fully develop the specialized markers that would make them incompatible with another person's blood system. Cord blood doesn't need to come from a sibling or a perfectly matched donor—it just needs to be in the ballpark in order to work.

"There's a woman somewhere out there who gave birth one day, and just decided to say, 'Yes, you can keep the medical waste,'" Julie says. "For Jenna it was no different than any other type of transfusion, procedurally. But this is what was going to save her life . . . It's unbelievable. This little bag of cells was going to save her life."

And it did. This clinical trial of a new cell-based medicine saved Jenna's life. By 2015 Jenna had been cancer free for three years, a sixth-grader who enjoyed singing and making up hip-hop dance routines. When she grows up, she wants to be a nurse.

This is how clinical trials should work. A patient who had progressed past what established medicine had to offer was able to try a promising experimental treatment that ended up saving her life. Unfortunately, it's not always this easy.

Some clinical trials don't work. When that happens, patients volunteering for these trials may suffer side effects without any benefit and, at the very least, lose precious time that could have been used trying other treatments.

"First, do no harm" is a guiding principle of medical ethics. Clinical trials ask doctors to gamble against this principle, to bet that the experimental treatment will do more good than harm. This can be an especially uncertain gamble in early trials, when the goal is not necessarily to benefit the individual patients but to make essential discoveries that could help other patients down the line—not to mention test the safety of the treatment.

The rules of clinical trials seek to ensure that researchers, doctors, and pharmaceutical companies are making ethical gambles and that patients understand the risks they are taking. Again, the chance for healing has to outweigh the chance for harm. Today, as new technologies lead to the development of new medicines and even new *kinds* of medicines, including those based on cells, the nature of these ethical challenges is changing at a pace never before seen. Understanding these changes requires a quick look back at the way trials have been run in the past, for better and for worse.

THE HISTORY OF CLINICAL TRIALS

In Judeo-Christian scripture, the Book of Daniel tells one of the earliest stories in which observers tried to discover the effects of a health intervention. In the story, Nebuchadnezzar, the king of Babylon, decrees that his people should eat only meat and drink only wine. A group of nobles prefers vegetables and eventually persuades Nebuchadnezzar to allow them to eat legumes for 10 days. At the end of these 10 days, the legume eaters seemed better nourished than the Babylonians who ate only meat, so Nebuchadnezzar added legumes to the menu. There was no mention what effect the exclusive dependence on wine had.

The eighteenth-century maritime physician James Lind, MD, is generally credited with overseeing the first scientific test to compare the effectiveness of medicines. In his *Treatise of the Scurvy* he writes, "On the 20th of May, 1747, I selected twelve patients in the scurvy, on board the *Salisbury* at sea. Their cases were as similar as I could have them. Two were ordered each a quart of cyder a day. Two others took twenty-five drops of elixir vitriol three times a day . . . Two others took two spoonfuls of vinegar three times a day . . . Two of the worst patients were put on a course of sea-water . . . Two others had each two oranges and one lemon given them every day . . . The two remaining patients, took . . . an electary recommended by a hospital surgeon . . . The consequence was, that the most sudden and visible good effects were perceived from the use of oranges and lemons."

Lind's story has a twist that presages the economic challenges of modern drug development. He had discovered that oranges and lemons were the best treatment for scurvy, but the fruits were so expensive that fifty years passed before the British Navy began using them consistently to keep their sailors healthy.

During the next two centuries, somewhat informal tests of new medicines, procedures, and devices continued, many with little oversight, questionable ethics, and methods that led to questionable results. For example, Lind wrote that "two of the worst" were treated with saltwater. If these sailors were among the sickest to begin with, who's to say it wasn't the advanced stage of their illness and not treatment with seawater that made them worse yet? Along these same lines, what if Lind had given citrus to the most promising patients? In that case, it could have been the natural course of recovery and not the oranges and lemons that led to the condition clearing up. Almost exactly two centuries after Lind, medical researchers wised up to the potential for "selection bias" and introduced the idea that researchers shouldn't be able to cherry-pick the patients who do or do not receive an experimental treatment. Today, random selection of which patients receive which treatments in a trial minimizes the chances that the results could be due to bias in choosing patients and therapies.

The first randomized-control trial was in 1946, when British doctors tested the antibiotic streptomycin as a treatment for tuberculosis. First, the trial had to clear a moral hurdle that has been the bane of many clinical trials. With strong evidence that antibiotics would indeed cure tuberculosis, how could doctors justify offering the treatment to some patients while withholding it from others?

In this case, circumstances provided the answer. During World War II, the United Kingdom had depended on the United States to supply antibiotics, so in 1946 the supply was extremely short. There simply was not enough streptomycin to go around, so some patients would necessarily receive the treatment while others would not. Randomizing the experimental and control populations would make for an objective test of the treatment, and the lottery-like system was the fairest way to decide who was treated.

In the modern era, researchers solve this moral dilemma of offering treatment to some but not to others by comparing new medicines against "standard of care." Instead of pitting a new drug against a placebo (in which case some patients could be saved while others would not be treated at all), many clinical trials add a new drug to the treatment that patients would get otherwise, ensuring that everyone receives the "best" treatment, while some just happen to get what might prove to be an added bonus.

When trials do compare a drug to a placebo, there are generally triggers that let researchers offer the drug to all patients on the trial if preliminary results make it seem especially effective.

At any rate, the 1946 British trial was a success, and streptomycin became the first drug approved to treat tuberculosis. (It still remains on the list for use if other, newer treatments fail.)

Another principle of research methodology that arose in the middle of the twentieth century is that clinical trials need to be "double blind," meaning that neither the researchers nor the patients know who has gotten a new treatment and who the placebo.

The famous physicist Richard P. Feynman, PhD, wrote, "The first principle is that you must not fool yourself—and you are the easiest person to fool." If a researcher knows that a patient has been given a treatment, the researcher may see improvements or side effects where none really exist. And when a patient knows he or she is trying a new drug, that knowledge alone can lead to improvements due to the placebo effect.

The modern gold standard of double-blind randomized controlled trials ensures that patients aren't grouped in ways that skew results, and that neither researchers nor patients find "results" in their wishful thinking.

The rules that now govern clinical trials are meant not only to ensure that the results are valid but also that patients are treated ethically. There is a historical reason for this. In 1932 the Public Health Service, working with the Tuskegee Institute in Alabama, started a unique piece of medical research. The "Tuskegee Study of Untreated Syphilis in the Negro Male" was intended to chart the natural course of untreated syphilis. Six hundred African American men, impoverished sharecroppers from Georgia, were enrolled after being promised treatment for "bad blood" in exchange for medical exams, meals, and burial insurance. The men were not, in fact, treated, even after 1947 when it was shown that penicillin could cure the disease. In fact, the "study" let hundreds of black men die of a disease that could be cured.

This unethical study is largely responsible for widespread mistrust and lack of participation in clinical trials by the African American community. The "Tuskegee Experiment," as it became known, is only one of many examples of medical misconduct, in and out of clinical trials.

In response to these cases of unethical human experimentation, in 1966 the World Medical Association adopted rules governing clinical trials. These rules for how human subjects are used in medical testing have been refined and updated over the years, and until now they have worked reasonably well.

CLINICAL TRIALS WITH TARGETED TREATMENTS

The sea of change in today's clinical trials is primarily in the fact that new targeted treatments don't—and aren't supposed to—work for everyone. Double-blind randomized-control trials, administered under versions of the 1966 guidelines, generally give a new drug to

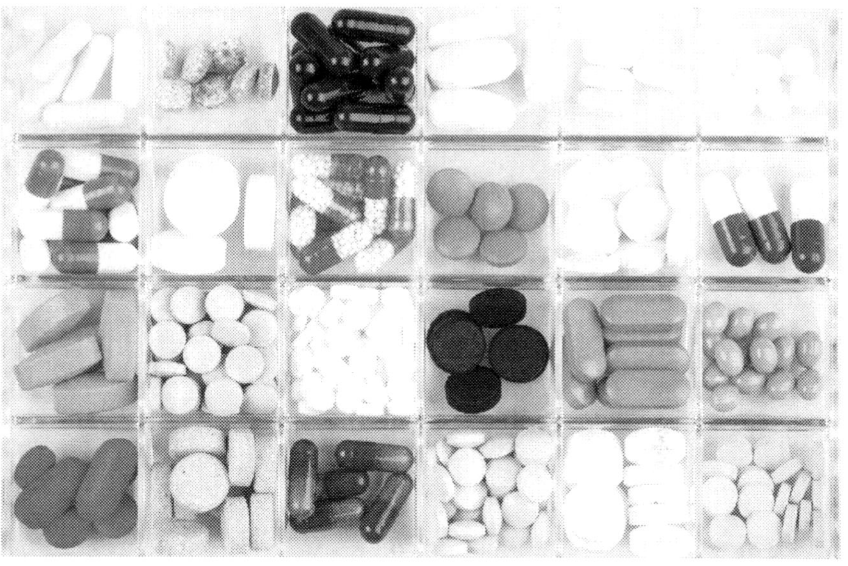

huge numbers of people. For example, during World War I, US Army medics noticed that exposure to mustard gas decimated soldiers' white blood cells. In the early 1940s, researchers Alfred Z. Gilman, Sr., PhD, and Louis S. Goodman, MD, started experimenting at Yale with mustard gas derivatives as medicines against non-Hodgkin's lymphoma, a cancer of the white blood cells. When the procedure proved promising, trials enrolled any willing patient with the disease, leading to the FDA approval of a slightly reengineered form of nitrogen mustard as the world's first chemotherapy.

This one-approach-treats-all method formed the basis for future trials, in and out of cancer. For example, in the 1980s the ISIS-3 study enrolled 41,299 heart-attack patients in a trial of medicines meant to break down blood clots. In the 1990s the Prostate, Lung, Colorectal and Ovarian (PLCO) Cancer Screening Trial tested over 150,000 people to see if screening could catch cancer early and thus lead to more cures.

But with today's drugs, one method does not treat all. We now know that, for example, not all breast cancers are the same disease. Some are driven by estrogen, some by progesterone, some by the HER2 gene, and some may even be driven by testosterone (an androgen), which is a known driver of prostate cancer and is just being recognized in a similar role in breast cancer. It would be useless to treat a breast cancer caused by HER2 with a drug that targets estrogen. In other words, unlike yesterday's chemotherapy trials against the entire population of breast cancer patients, new genetically targeted treatments are tested only against the breast cancer patients whose tumors hold a particular mutation.

Testing a targeted treatment against the general population could make the drug look ineffective when, in fact, it might be highly effective against the subset of people whose disease is specifically sensitive to it.

Therefore, the smaller the subset of patients, the more precisely a trial must identify the people likely to benefit while excluding those who won't. For example, 70 percent of breast cancers are estrogen positive. Testing a new antiestrogen against everyone with breast cancer would still have the potential to work with 70 percent of patients. On the other hand, only about 3 to 5 percent of lung cancers are caused by the gene translocation called ALK-EML4. If the drug crizotinib (XALKORI) had been tested against all lung cancer patients, it would *not* have worked for about 95 percent of patients and the drug would have appeared ineffective. And the last thing a pharmaceutical company wants is for its drug to appear ineffective.

Today, targeted treatments are not pushed into clinical trials alone but with companion tests meant to identify who is likely to benefit. The question now is not only whether a drug is effective, but in which patients will it be most effective.

ETHICAL CHALLENGES OF MODERN CLINICAL TRIALS

Of course, this intense need to show a drug's effectiveness is one thing that can drag modern clinical trials into murky ethical waters. Drug companies need to show that their medicines save lives. The pharmaceutical industry, doctors, and patients all have an incentive to work together to put a drug's best foot forward. Everyone wants to give the drug to patients who will benefit.

But what about Dr. James Lind's scurvy trial? Specifically, what about the sickest two crewmembers, who were prescribed a course of seawater? If they had been given lemons (vitamin C), would it really have made a difference? Maybe, maybe not. In other words, testing new drugs on patients who are beyond help may not be the best trial design, because the experimental drug might have a greater effect on people who are less sick.

Sometimes we can see this happening in clinical cancer trials. Many trials are conducted first with patients whose disease has spread and who no longer respond to available therapies. Because these patients are in the most desperate need of treatments, the testing of new drugs faces fewer regulatory and ethical hurdles. The same is true in testing against diseases such as ALS. The more desperate the disease, the easier it is to ethically justify possible treatments.

However, many clinical cancer trials now exclude patients whose disease has metastasized to their brains, and this is where the needs of patients and drug developers might diverge, at least in the short term. There are some new treatments, especially small molecules, that can cross the blood-brain barrier to attack cancer where it hides in the brain, as discussed in chapter four. But chances for dramatic improvement at this point are slim. No matter how well a drug works with most patients, the drug will likely fail in patients with metastatic cancer in the brain. So drug companies withhold some new medicines from these patients, even though there's an outside chance that these medicines might work. Excluding the most desperate cases can make the drug look statistically more effective.

However, this ethical challenge has provided another opportunity for clinical trials to evolve. In addition to testing drugs with the right patients, innovative clinical trials are learning to report their results by segment—describing which sets of patients benefited more and less. If the drug is ineffective in the sickest patients, the trial designers can still see how it fares in the people they most expect to help.

This tiny change has enormous implications. Clinical trials that lump together the results of all patients must exclude the sickest patients or risk making the drug look like a flop. Trials that report results for the sickest patients separately can offer experimental medicines to patients understood to be long shots.

CLINICAL TRIALS IN SMALL PATIENT POPULATIONS

However, microtargeting patients and then microreporting results brings a new challenge—finding enough patients to test these new drugs at all. This is especially true with rare diseases.

We discussed Duchenne muscular dystrophy and other rare disease in chapter ten. The Centers for Disease Control and Prevention estimate that in 2007, of 2.37 million American boys aged 5 to 24, only 349 lived with Duchenne. That's an average of only seven per

The challenge when a disease affects few people is finding enough volunteers to test treatment and also to make the treatment profitable enough to develop.

state. What are the chances that any single research center could enroll enough of these boys in a clinical trial to test a new treatment? The same is true for subtypes of larger diseases. It may have been easy to enroll enough patients to make conclusions about chemotherapies for more than 200,000 patients diagnosed with lung cancer each year in the United States, but it's much more difficult to gather enough people to test therapies that target ALK+ lung cancer (about 6,500 cases per year) or now ROS1+ lung cancer (about 4,500 cases per year).

The reality of microtargeted patient populations has forced a paradigm shift in clinical trial design. First, many centers are banding together to pool their clinical trial data, as with consortia like ORIEN (see chapter fifteen). If one center enrolls only 7 patients per year in a trial, then ten centers may be able to enroll 70. Second, new laws have relaxed clinical trial requirements in ways that can help people with rare diseases take advantage of experimental treatments.

One important change was the 2012 passage of the Food and Drug Administration Safety and Innovation Act. Until then, FDA approval depended exclusively on a treatment's being shown to improve patients' conditions in clinical trials. But for rare diseases like Duchenne or for small genetically targeted subgroups of a disease, human trials may be impractical or even impossible. You might never be able to collect enough patients for a trial, or it could take years and years to run a trial, during which time patients would continue to suffer.

For drugs that satisfy an unmet need in a serious condition, the Safety and Innovation Act now allows for drugs to be measured according to improvements in "surrogate markers" rather than overall patient improvement. Surrogate markers are laboratory measurements or physical signs used as a substitute for overall clinical outcomes. For Duchenne, the surrogate marker was increased dystrophin production. You will remember that a lack of this muscle-repair protein causes Duchenne. Thus, a drug's influence on dystrophin is a reasonable marker of its effect.

In fact, it was Pat Furlong, the woman you read about in chapter ten whose sons died from the condition, who crafted the new guidelines for drug development in Duchenne, along with scientific experts associated with the organization she founded, Parent Project Muscular Dystrophy. In 2015 her suggestions became rules, as the FDA stated that under the Safety and Innovation Act, dystrophin production alone could be used to accelerate new drugs toward approval.

A somewhat newer approach to testing therapies for rare or difficult-to-treat cancers has been dubbed "basket studies" by researchers at New York's Memorial Sloan Kettering Cancer Center. While traditional clinical trials focus on a particular cancer type, basket studies are concentrated on a specific mutation found in the tumor, regardless of where the cancer originated.

Published results using this approach demonstrate that researchers can design trials based on genomics as opposed to the site of origin of the cancer. This means that a particular clinical trial may actually include several different types of cancer that arise from the same mutation. Even more exciting is that the ability to characterize a tumor based on its genetics and then match it with a drug known to target that mutation means that there's a reasonable expectation that a given drug may have a significant clinical benefit, even in an early phase clinical trial.

Cellular medicines are not chemotherapy. They may only be useful with small targeted groups of patients. In other words, their goal may not be incremental improvements in many people—as with drugs that reduce cholesterol in millions of Americans—but dramatic cures in a very few. In the past, clinical trials haven't been designed to discover these very sharp needles in very large haystacks. In the future, they will have to be.

Today we are at the epicenter of this shift, transitioning from an era of a few widely used medicines to an era defined by thousands of highly specialized ones. Our clinical trials, like these medicines themselves, are just starting to find their new form.

CHAPTER EIGHTEEN

CELLS WILL BE THE DRUGS OF THE FUTURE

Prediction is very difficult, especially about the future.

—NIELS BOHR, PHD

In 2006 MIT's *Technology Review* offered a $20,000 prize to anyone who could disprove the proposals of Aubrey de Grey, PhD, to radically extend the human life span. The prize went unclaimed. This is despite the fact that many mainstream scientists consider de Grey's ideas—and the man himself—a little sideways of serious science.

Just consider the belief that underlies his thinking: to de Grey, aging is a disease that can be prevented and treated. By taking advantage of new cellular therapies, de Grey says, we can repair the damage incurred

by a lifetime of molecular and cellular destruction and live longer than we ever believed possible, perhaps hundreds of years longer.

His strategy lies in counteracting what he sees as the seven causes of aging, which he calls the "seven deadly things." These are:

- The inability of muscles, the heart, and the brain to keep pace with cell loss.
- Uncontrolled cell division (cancer).
- The accumulation of toxins produced by damaged mitochondria (the body's cellular power plants).
- The buildup of malfunctioning cells.
- The stiffening of tissues (such as those in our arteries).
- The accumulation of extracellular and intracellular "junk."

When left unchecked, de Grey believes this accumulated cellular damage leads to a host of ailments, from heart disease, arthritis, and type 2 diabetes to Alzheimer's disease and a variety of cancers. De Grey's strategy is to pick off these seven deadlies one by one.

De Grey also believes that existing cellular medicines and ones that are already showing promise in development are well on their way to making inroads against these seven drivers of aging. These therapies include stem cells that restore cells in organs that are losing them, enzyme therapies that remove molecular garbage from inside cells, immunotherapies that remove unwanted detritus from the space between cells, and drugs that will break the chemical linkages that stiffen arteries.

In 2009 de Grey cofounded the SENS (Strategies for Engineered Negligible Senescence) Research Foundation, based in Mountain View, California, to directly gather and fund scientists to pursue these challenges. In a 2013 interview with *Newsweek*, de Grey noted that the SENS Foundation's work is "not about slowing down aging at all, but actually reversing it—repairing the accumulation of molecular and cellular damage that builds up throughout life as a side effect of the body's normal operation."

De Grey believes that if doctors can make these repairs, there is a very good chance that they could rejuvenate people so that "we will be able to take them back to being biologically 30 or 40." He and his team hope to be able to make these repairs "repeatedly, so that people can stay genuinely youthful for a lot longer and reduce the risk factors that lead to dying." De Grey's timeline is thirty years. In an interview with *Motherboard*, he said, "These therapies are going to be good enough to take middle age people, say people aged sixty, and rejuvenate them thoroughly enough so they won't be biologically sixty again until they are chronologically ninety. That means we have essentially bought thirty years of time to figure out how to re-rejuvenate them when they are chronologically ninety so they won't be biologically sixty for a third time until they are 120 or 150. I believe that thirty years is going to be very easily enough time to do that."

LEGITIMACY IN DOLLARS

With any futuristic project like radical life extension, there are visionaries like Dr. Aubrey de Grey thinking so far ahead of current science that it's hard to see how we might get from here to the future they predict. However, while mainstream scientists might not share de Grey's radical, beyond-the-fringe optimism, he has nonetheless stretched mainstream thinking in that direction. It's not only SENS scientists who are working hard on solutions for each of de Grey's seven aspects of aging, and many have been able to secure perhaps the most important stamp of legitimacy, funding from outside sources.

Today, major drug companies have started investing in regenerative medicine. In 2014, Google launched Calico Labs and partnered with the pharmaceutical giant AbbVie on a $500 million venture focused on antiaging research. Also joining in the discovery, development, and commercialization of new therapies for age-related diseases are Oracle's Larry Ellison, billionaire investor Peter Thiel, Amazon founder Jeff Bezos, and Napster pioneer and early Facebook president Sean Parker. That's not to mention Facebook founder Mark Zuckerberg and his wife, Priscilla Chan, MD, who have pledged $3 billion to rid the world of major diseases during the lifetime of their daughter born in 2015. Each has directed millions or billions of dollars in personal funds toward cell-based therapies designed to slow the effects of aging and treat the diseases that have come with age through all human history.

These megaphilanthropists have engaged leading scientists in the pursuit of cellular medicines. For example, heading the Parker Institute for Cancer Immunotherapy (a collaboration of six leading American cancer centers, launched in 2016) is Jeffrey Bluestone, PhD, whose groundbreaking research focuses on treatments based on T cells trained to attack genetic targets.

Bluestone also works on the other side of T-cell activation to ensure that these cells do not improperly attack the body's healthy tissues. His work has led to the development of multiple therapies that promote immune

Dr. Jeffrey Bluestone, president of the Parker Institute for Cancer Immunotherapy.

tolerance, including CTLA4Ig (abatacept, brand name ORENCIA), used in the treatment of rheumatoid arthritis; belatacept (NULOJIX), the first FDA-approved drug targeting T-cell costimulation to treat autoimmune disease and organ transplant rejection; a novel anti-human-CD3 antibody being developed to treat type 1 diabetes; and the first CTLA-4 antagonist drugs approved for the treatment of metastatic cancer.

In his recent research, Bluestone has switched from manipulating the immune system's attack signals, to working with the stand-down signals that can deactivate the immune system.

This work focuses on the critical role of regulatory T cells (Tregs) in autoimmune diseases such as type 1 diabetes, lupus, and multiple sclerosis. As we have seen in chapter seven, Tregs can calm the aggression of T-cell attacks. While the present science of immunotherapies has to do with turning the immune system on to destroy viruses and cancer cells, the future may be more about learning to turn off the system to treat autoimmune diseases.

As philanthropists are choosing to bet their money on regenerative medicine, many researchers from top universities are betting their careers on it. Academic research is incredibly competitive, with success depending on a scientist's ability to predict what is next and then deliver on the promise. The fact that cellular strategies are in Ivy League labs not only promises to push the research forward, but demonstrates that these strategies are no longer on the fringe but have been embraced by mainstream science.

HARVARD EXPERTS FOCUSING ON REGENERATIVE MEDICINE

A decade ago, when Lawrence H. Summers, PhD, delivered his last commencement address as the president of Harvard University, he pointed to a rosy future in which cancer, diabetes, and Alzheimer's would be treated with therapies based on stem cells. Summers, who is now the Charles W. Eliot University Professor at Harvard, has come to believe that he severely underestimated the number of conditions that cellular therapies will be able to treat.

Addressing the attendees at the Science and Business of Regenerative Medicine symposium at Harvard in 2016, Summers said, "I did not appreciate what subsequent research has demonstrated [as] stem cells' capacity to address sickle cell anemia, to grow new heart cells and repair aging hearts, to replace tendons and ligaments, to address blood cancers . . . and to address certain forms of blindness that have been cured by stem cell transplants."

Summers envisions a future in which cellular therapies are no longer a niche technique being tested against a few diseases but a major branch of biomedical research and treatment. Even today, doctors are using cellular strategies to treat more diseases, but are using them earlier in the treatment process. There is growing confidence that these cell-based medicines already work as well or better than earlier generations of therapies.

Mark Fishman, MD, one of Dr. Lawrence Summers' colleagues at Harvard, echoed this optimism about the ongoing and future possibilities of cellular therapies. Dr. Fishman, who is a professor of stem cell and regenerative biology, believes that regenerative medicine will prove to be the third big wave of transformative medicine, following the control of infectious disease in the last century and today's continually promising work to fight cancer with novel immunotherapies.

Fishman suggests that the wave of regenerative medicine will be composed of two distinct approaches. The first is a "spare parts" approach focused on developing cells, tissues, and organs to replace those diseased or damaged in the body. The second focuses on mobilizing the body's own regenerative capacity—to harness, restart, support, or engineer the body's ability to replace and repair its own tissues. This ability exists early in human development and in adult creatures such as zebrafish and salamanders. This wouldn't require creating a new biological ability, but merely tapping into nature's preexisting programs that happen to be hidden or suppressed in adult humans.

Fishman sees this new wave of cell-based therapies aimed at increasing the health span as inevitable, in part because there is no other

choice. "Regenerative medicine has to be the next great frontier," says Fishman, "The aging of the global population increases the urgency of finding treatments to keep people healthy longer. We have no choice; we have to make this work."

THE FUTURE OF CELL-BASED THERAPIES

The field of cell-based medicine is growing so fast in so many directions that it's difficult for anyone to keep up with all the developments. Every day, promising treatments fail to pan out, while ideas that seemed like pipe dreams unexpectedly blossom into elegant new ways to attack disease. Such was the case for immunotherapies. Twenty years ago, immunotherapy was largely a fringe science and now it is a major category of treatments against a spectrum of diseases.

In other branches of cell therapy, new strategies are starting to emerge from basic research into the light of day. One example out of many is nerve growth factor (NGF) in Alzheimer's disease.

NGF was discovered in the 1940s by Rita Levi-Montalcini, MD, a neurologist who showed that this small protein supports sensory neurons during early development of the nervous system. In 1986 she was awarded the Nobel Prize for her discoveries. Since then, other scientists have shown that NGF also promotes the survival of acetylcholine-producing cells in the basal forebrain, cells that are eventually eliminated during the course of Alzheimer's. But until recently, it has been impossible to take the obvious next step: if acetylcholine-producing cells are lost in Alzheimer's and NGF supports these cells, why not add more NGF?

As with many of these frustrating dilemmas—where a clear understanding of the problem suggests that there *should* be an easy solution—the answer lies in the deep and complex science of "druggable targets." It can be extraordinarily difficult to decrease the amount of a specific protein in the body, and even harder to increase it. This is especially true when the target is in the brain, where many molecules that could be administered orally or intravenously are too big to cross the blood-brain barrier.

This is why the theoretically simple idea of adding more NGF to the Alzheimer's patient's brain seemed to be on the same horizon as a human mission to Mars. The impossibility of delivering NGF to the brain led most pharmaceutical companies and researchers to shift their focus elsewhere. Many have spent the last twenty years trying to develop drugs to remove beta-amyloid plaque from the brains of people with Alzheimer's. So far, nothing has worked.

During this time, Mark Tuszynski, MD, PhD, director of the Translational Neuroscience Institute at the University of California, San Diego, and his team gambled on what seemed like a long shot—continuing to push forward with NGF. Tuszynski hoped to use gene therapy to boost the brain's capacity to heal itself. Eventually, he and his colleagues developed and finally settled on a futuristic approach that entered a small clinical trial.

First, his team took skin biopsies from patients' backs. Tuszynski then isolated a special kind of connective-tissue cell called a fibroblast and genetically modified the fibroblasts in the samples to express the genes that make NGF. In a surgical procedure, he then directly

implanted those cells into the basal forebrains of 10 patients with Alzheimer's disease. This area maintains memory, attention, and behavior, and it is usually affected early in the development of Alzheimer's. It is also a part of the brain where NGF seems especially active in maintaining the health of cells responsible for those functions.

As in most early clinical trials, the purpose was not so much to save lives as to show that the biology was promising. As expected, these 10 patients lived an average of five more years, yet autopsies of their brains showed real signs of improvement.

In a recent article in *JAMA Neurology*, Tuszynski and his team reported that, in the brains of all 10 patients, cells that had been in the process of dying due to Alzheimer's had instead regrown axons in regions where NGF had been delivered. This was a major development not only because the therapy worked but because science had finally found a way to use NGF as a drug. The secret was not a new pill but the creative use of cells as literal medicine.

In a phase II trial at the University of California, San Diego, and nine other sites around the country, researchers are now testing to see if this technique can improve the memory and cognition problems caused by Alzheimer's. Dr. Tuszynski told the *Washington Post* that if the results continue to look promising, a phase III trial could pave the way for patients to "have a surgical procedure of three to four hours and have lifelong protection, without having to take a drug every day. If you know that you have a disease that robs you of the essence of your intellect, will you undergo a three-hour operation? Pardon the expression, but that seems like a no-brainer."

THE FUTURE OF KIDNEY TRANSPLANTS

Certain kinds of cells may work individually, acting like tiny factories or pharmacies or police officers or peacekeepers. Cells working together can act like organs.

According to the United Network for Organ Sharing, there are currently 93,000 people on the kidney transplant list. It generally takes about five years for a person on the list to be matched with a donor's kidney—although the average wait in some states is as much as ten years. With only 20,000 kidneys available annually, scientists have been racing to find an alternative.

It looks increasingly likely that the alternative may be engineered cells working together to form a replacement kidney. The device, pioneered by researchers from the University of California, San Francisco, and Vanderbilt University Medical Center, is about the size of a coffee cup and is powered by a bioreactor driven by blood pressure—no need for a pump or electrical power. Installed in place of a malfunctioning kidney, the device, built from kidney cells, would screen out toxins, salts, and some small molecules from the blood, mimicking the natural actions of the kidney. In the process, water is also reabsorbed back into the body, where it will eventually be directed to the bladder for excretion. Human trials start in 2017 with researchers hoping that the surgically implanted device will function as a normal kidney and keep patients off dialysis.

THE FUTURE OF HEALING BONES

Of an average six million broken bones in the United States each year, about 300,000 will be slow to heal or will not heal at all with traditional methods. For years, scientists have tried to cajole stem cells into aiding this process. One problem that has frustrated these attempts is the fact that stem cells tend to either die off or migrate away from the repair site. Basically, stem cells don't stick around long enough to influence the pace of repair.

A new technique now solves this problem, opening the door for the future use of stem cells in bone repair. As explained in a paper published in the journal *Acta Biomaterialia*, before stem cells are injected into the site of bone repair, they are first encased in polymers called hydrogels. As the word implies, hydrogels hold water, which is necessary for the survival of the stem cells. These hydrogels degrade over time, disappearing before the body interprets them as foreign bodies and begins a defense response that could compromise the healing process.

The research team, headed by Danielle Benoit, PhD, assistant professor of biomedical engineering at the University of Rochester, hoped these hydrogels would allow the stem cells to finish the job of initiating repairs, then leave before overstaying their welcome.

Benoit and her team tested this strategy in mouse bone grafts. First, they removed all living cells from the new bone to be engrafted, so that any healing could only be the result of stem cells. Then, they attached the bone graft and injected hydrogel-coated stem cells as a type of healing glue at the point of attachment.

What they found is promising. Encased in hydrogel capsules, these stem cells stayed put. As with the new technique for drugging NGF in the brain, the fact that the team could keep stem cells localized at the site of bone repair solves one distinct, essential problem on the path to a viable therapy.

"Our success opens the door for many—and more complicated—types of bone repair," says Benoit. The team is pushing these hydrogel stem cell systems toward FDA approval.

LOOKING AHEAD

In the coming years, cellular therapies will be the home of both hope and hype. There will be overpromising blind alleys, wrong turns, and occasionally even discredited research. Science is a human endeavor, and sometimes it goes wrong. However, we have realized enough promise in cell therapies to have passed an important tipping point. We now know that cellular therapy can heal or replace tissues and organs damaged by disease, trauma, or age. We know that despite the

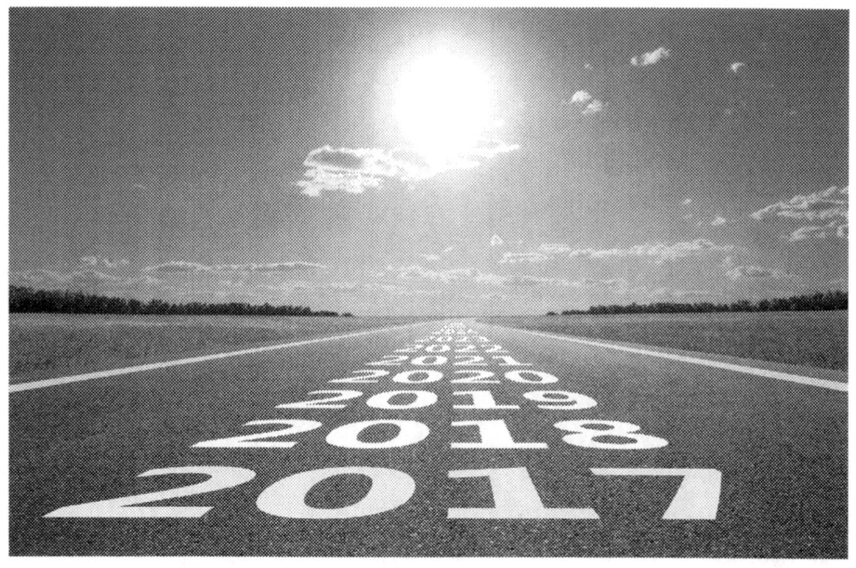

inevitable dead ends, there will also be breakthroughs. The question is not whether cell therapies will work but how and when.

The answers will come from scientists hard at work in their laboratories uncovering new knowledge about genes and stem cells and proteins and our ability to control their actions. Some of this knowledge will move from laboratories to clinics, becoming cell-based treatments for many diseases affecting millions of people. Other discoveries will change the *ways* in which we treat disease, with applications that could be used across entire classes of disease, or to pick off some of the rarest but most devastating human conditions.

Realizing this future requires perseverance, funding, and hard work. Each of us is a collection of over thirty trillion human cells, and we are entering an age of medicine in which those very cells will be the drugs that keep us alive longer and in better health than ever before.

ACKNOWLEDGMENTS

While healing comes from within, strength comes from those around us. I am grateful to my coauthor and friend Dr. Max Gomez, and thankful for the love and support of Matthew Henninger and my parents, Norma and Gordon Smith.

I know you will all join me in saluting the true heroes: the patients, doctors, researchers, and institutions who are bringing hope to the many people around the world who are suffering with debilitating diseases.

A special thank you to Dr. W. E. Bosarge, who has generously supported our vision over the past seven years, and his unwavering devotion to support regenerative medicine, personalized medicine, and cancer research. I am forever grateful to my mentors—Dr. Charles Greer, who introduced me to stem cells, and Dr. David Nash, who taught me the business principles of medicine.

This book could not have been completed without the help of Gerald Couzens (may he rest in peace), Garth Sundem, Svetlana Izrailova-Danov, Sheree Bykofsky, and the team at BenBella Books, and we are thankful for each of their contributions.

—ROBIN

It is impossible to read of the courage, perseverance, and innovation we recount here and not be awed by the patients, families, and researchers who have brought us to this inflection point in medicine. I wish to express my sincere gratitude to them as well as to the people in my life who enabled me to tell these stories.

I am very thankful for the counsel of my friend and colleague, Dr. Robin Smith; for my late parents, who always supported and encouraged my nontraditional career path; for the patience of my children, Katie and Max IV; and for the love and support of Wendy Dessy.

I will also be forever grateful to my friend and mentor, Dr. Inglis Miller, who guided me in my development as a scientist and an individual.

—MAX

PHOTO CREDITS

Page 74, Lasker Foundation logo used with permission

Rights to the following images were purchased from iStockphoto.com:

CHAPTER ONE

Page 3, ©iStock.com/Grafissimo

Page 4, ©iStock.com/ttsz (source image)

Page 5, ©iStock.com/ttsz (source image)

Page 8, ©iStock.com/vshivkova (bottom, source image)

Page 9, ©iStock.com/blueringmedia (source image)

Page 13, ©iStock.com/blueringmedia

CHAPTER TWO

Page 28, ©iStock.com/Fly_dragonfly

CHAPTER THREE

Page 36, ©iStock.com/haydenbird

Page 37, ©iStock.com/colematt (source image)

Page 38, ©iStock.com/muuraa (source image)

Page 40, ©iStock.com/baranozdemir (source image)

Page 42, ©iStock.com/selvanegra

Page 44, ©iStock.com/blueringmedia

Page 47, ©iStock.com/OcusFocus

CHAPTER FOUR

Page 58, ©iStock.com/Terriana, ©iStock.com/royaltystockphoto, ©iStock.com/Kubkoo, ©iStock.com/MicrovOne (source images)

Page 63, ©iStock.com/Reptile8488

CHAPTER FIVE

Page 70, ©iStock.com/ttsz (source image)

CHAPTER SIX

Page 86, ©iStock.com/Hailshadow

Page 88, ©iStock.com/ttsz

Page 92, ©iStock.com/blueringmedia (source image)

Page 94, ©iStock.com/sleddogtwo (source image)

Page 97, ©iStock.com/ttsz

CHAPTER SEVEN

Page 103, ©iStock.com/Kora_ra (source images)

Page 104, ©iStock.com/AzmanL

Page 106, ©iStock.com/DIGIcal

Page 107, ©iStock.com/AlexRaths

Page 113, ©iStock.com/TLFurrer

Page 117, ©iStock.com/ttsz (source image)

Page 119, ©iStock.com/Val_Iva, ©iStock.com/Marchiez, ©iStock.com/setory, ©iStock.com/TheresaTibbetts, ©iStock.com/Epine_art, ©iStock.com/miu_miu, ©iStock.com/

MegaShabanov, ©iStock.com/Anna_zabella, ©iStock.com/olegtoka, ©iStock.com/LplusD, ©iStock.com/venimo, ©iStock.com/Fafarumba, and ©iStock.com/DenisDubrovin (source images)

CHAPTER EIGHT

Page 122, ©iStock.com/7postman

CHAPTER NINE

Page 139, ©iStock.com/SilverV

Page 141, ©iStock.com/Perception7 (source images)

Page 146, ©iStock.com/danchooalex, ©iStock.com/279photo, ©iStock.com/EJ_Zet

CHAPTER TEN

Page 155, ©iStock.com/Tigatelu, ©iStock.com/lvcandy, and ©iStock.com/ttsz (source images)

Page 156, ©iStock.com/FatCamera

Page 160, ©iStock.com/vchal

CHAPTER ELEVEN

Page 166, ©iStock.com/Dr_Microbe

Page 170, ©iStock.com/HAYKIRDI

Page 173, ©iStock.com/Pogonici

CHAPTER TWELVE

Page 177, ©iStock.com/from2015 and ©iStock.com/JuliarStudio (source images)

Page 181, ©iStock.com/Lonely__

Page 185, ©iStock.com/AlexRaths

CHAPTER THIRTEEN

Page 189, ©iStock.com/luismmolina

Page 192, ©iStock.com/kchungtw

Page 193, ©iStock.com/ttsz

Page 195, © iStock.com/ttsz (source images)

Page 198, ©iStock.com/miralex

Page 201, ©iStock.com/luismmolina

Page 206, ©iStock.com/ttsz (source image)

Page 210, ©iStock.com/baona

CHAPTER FOURTEEN

Page 212, ©iStock.com/KeithBishop

Page 215, ©iStock.com/vandervelden

Page 217, ©iStock.com/pacopole

Page 220, ©iStock.com/bubaone

CHAPTER FIFTEEN

Page 225, ©iStock.com/Maxiphoto

Page 229, ©iStock.com/semnic

Page 232, ©iStock.com/gilaxia

CHAPTER SIXTEEN

Page 241, ©iStock.com/weerapatkiatdumrong

Page 254, ©iStock.com/artisteer

CHAPTER SEVENTEEN

Page 257, ©iStock.com/alexskopje

Page 259, ©iStock.com/fstop123

Page 263, ©iStock.com/Viperfzk

Page 266, ©iStock.com/elenabs

CHAPTER EIGHTEEN

Page 270, ©iStock.com/sportpoint

Page 275, ©iStock.com/AleksandarNakic

Page 277, ©iStock.com/cosmin4000

Page 281, ©iStock.com/honglouwawa

INDEX

Note: *Italic* page numbers indicate illustrations or photographs

A

AACR Distinguished Public Service Award, 242
Aarskog syndrome, 151
abatacept (ORENCIA), 273
AbbVie, 272
Aboody, Karen, 62–63
Abramson Cancer Center (University of Pennsylvania), 80–83
academia, philanthropy supporting medical research in, 253
acetylcholine, 100, 229
Acta Biomaterialia, 279
ACTEMRA (tocilizumab), 128
acute anaphylaxis, 105
acute lymphoblastic leukemia (ALL)
    CAR-T therapy for, 126–132
    GLEEVEC for treatment of, 143
acute myeloid leukemia (AML) cells, 168–169
ADAM10 receptor, 172
adoptive T cell strategies, for SCID and HIV treatment, 133
adult stem cells, 7, 9
African Americans, 262
age, cancer risk and, 219–222
aging, halting/reversing. *see* life expectancy extension
Aging Cell, 203
aging tissue, regeneration of. *see* tissue regeneration
AIDS epidemic, 190
Albert Einstein College of Medicine, 190
alcohol consumption, and life expectancy, 218
Alex (boy with severe allergies), 106–109
Ali, Muhammad, *51,* 163
ALK-EML4 gene, 264
ALK gene, 167, 227
ALK+ lung cancer, 267
ALL. *see* acute lymphoblastic leukemia
Allen, Woody, 185
allergies, 102–115
    asthma, 109–111
    immunotherapy for, 105–109
    major food allergens, *103*
    and microbiome, 112–115
    philanthropy supporting research on, 251
    skin testing, *107*
    Tregs and, 109–111
Allison, James P., 72, *75,* 75–77
Allogeneic Human Mesenchymal Stem Cell Infusion versus Placebo in Patients with Alzheimer's Disease study, 43–45
alpha hemolysin, 172
ALS. *see* amyotrophic lateral sclerosis
ALS Association, 46
Alt, Eckhard, 7
Alzheimer's Association, 204
Alzheimer's disease, xv
    first human stem cell trial, 43–45
    NGF for treatment of, 276–278
    philanthropy supporting research on, 250
    senolytic drugs for treatment of, 204–205, 207
    stem cell treatment, 40–45, *42, 44*
Amazon, 272
American Academy of Microbiology, 112
American Cancer Society, 125
American Federation for Aging Research, 191
American Heart Association, 198–199
American Journal of the Medical Sciences, 68
American Neurological Association, 49
American Society for Blood and Marrow Transplantation (ASBMT), 60
AML (acute myeloid leukemia) cells, 168–169
Amorcyte Therapeutics, 18
amyloid plaques, *42,* 42–44
amyotrophic lateral sclerosis (ALS), xv, 265
    about, 45–47
    stem cell research, 47–50
angelMD, 254
Anna Karenina (Tolstoy), 189

Annals of Nutrition and Metabolism, 116
antiaging therapies, 200, 202–204. *see also* life expectancy extension
antibiotics
and *C. diff.* infection, 113
and development of allergies, 115
and gut bacteria, 164
supply of, in 1940s, 261
anti-CTLA-4 drugs, 75, 76. *see also* ipilimumab
antigens, treating asthma with exposure to, 111
antigen-specific therapy, for asthma, 111
antihistamine medications, 106
antirejection drugs, 199
AP20187 (compound), 202, 203
apoptosis, 200
Archives of Internal Medicine, 212
Archives of Neurology, 214–215
Array BioPharma, 231
ASBMT (American Society for Blood and Marrow Transplantation), 60
Asian carp, 180
aspirin, 230
Association of University Technology Managers, 240
asthma
bone marrow transplant and, 104, 105
philanthropy supporting research on, 251
Treg cells and, 109–111
Atala, Anthony, 20–27
atezolizumab (TECENTRIQ), 121–122
ATP, 208
Austin, Thomas, 180
Australia, 180, 247
Autism Speaks, 250
autism spectrum disorders, philanthropy supporting research on, 250
autoimmune diseases, 85–101
allergies, 102–115
gluten intolerance/celiac disease, 116–118, *117, 119*
multiple sclerosis, 91–96, *92*
myasthenia gravis, 91, 98–101
philanthropy supporting research on, 251, 273
prevalence of, 90–91
type 1 diabetes, 86–90
autologous stem cell transplants
as multiple sclerosis treatment, 94–96
as myasthenia gravis treatment, 98–101
nonmyeloablative, 94–96
resetting the immune system with, 96–98
avascular necrosis (AVN), 13–15
Avastin (bevacizumab), 84, 121–122, 167
AZD6244, 231
azolla, 255

B

baboons, 197
Baker, Darren, 201–202
Balo disease, 151
Barnard, Christiaan, 197–198
Baruch, Bernard, 210
Barzilai, Nir, 190–193
basket studies, 268
Baumel, Bernard S., 43–45
Baxter Laboratory for Stem Cell Biology, 196
Bayh-Dole Act, 240
B-cell leukemia, CAR-T therapy for treatment of, 127
B cells, 69, 104, 105
BCL2 gene, 168
BCR-ABL gene, 165, 166, 203
belatacept (NULOJIX), 273
Bell, Alexander Graham, 189
Belvin, Sharon, 75–77
Bennett, Tony, 211
Benoit, Danielle, 280
beta cells, pancreatic, 87–89, 134
beta thalassmia, 146, 176–177
BeTheMatch.org, 57
bevacizumab (Avastin), 84, 121–122, 167
Beygi, Roxane, 91, 96, 101
Bezos, Jeff, 272
Biden, Joe, 227
big data
and disease prevention, 234–235
and drug design, 232–235
and drug libraries, 230–232
and drug repositioning, 227–230
and medical research, 224–236
ORIEN and other data-sharing collaborations, 226–227
statistics and sample size, 224–226
binding pocket, protein, 169
biofilm, 25
biomedical sector, 179
"bionic pancreas" project, 247–248
bioprinting
bladder, 21–22, *22*
ear, *23*
Biostage, 30, 31
bispecific T-cell engagers (BiTEs), 125
bladder, 20–22, *22*
Blau, Helen M., 196–197
blindness, LCA as cause of, 139–140
Blood (journal), 104
blood-brain barrier, 61, 62, 265
Bloomberg TV, 210
Bluestone, Jeffrey, 272–273, *273*
blue whales, cancer risk for, 219, 221
BMAC (bone marrow aspirate concentrate), 15
BMI (body mass index), 216
Boca Raton Regional Hospital, 250
BOC Sciences, 231
body mass index (BMI), 216
body parts. *see* organs, laboratory-grown
Bohr, Niels, 269
bone marrow
hematopoietic stem cells and, 10
as source of stem cells for ALS treatment, 48
bone marrow aspirate concentrate (BMAC), 15
bone marrow transplants
allergies following, 102–105
for blood disorders, 114
Jenna Gibson and, 257–258
and GVHD, 55
immunotherapy with, 132–133
stem cell treatment for GVHD, 56–60
bone repair, future of, 279–280
Bosarge, W. E., 242, *242*

Boston Children's Hospital, 85–87
Boston Red Sox, 224
Boston University, 248
BQU57 (drug), 235
Bradbury, Ray, 180–181
Brady, James Buchanan ("Diamond Jim"), 237–239
BRAF mutation, combined therapy for cancer caused by, 124–125
brain, 35–53
  ALS treatment, 45–50
  Alzheimer's disease, 40–45, 276–278
  cancers in, 61–64, 265
  cell senescence in, 204–207
  Parkinson's disease, 50–52
  physical exercise and health of, 214–215
  social engagement's effect on, 216–217
  stroke, 35–39, *37, 38*
BRAIN Initiative, 227
BrainStorm Cell Therapeutics, 48
Brazil, 170
BRCA gene, 140
breast cancer, 84
  and alcohol consumption, 218
  clinical trials of treatment for, 263–264
  oncogene inhibitor for treatment of, 167
brentuximab, 84
Broad, Eli, 1
Brown University, 208
Bryant, Kobe, 11–12
"bubble boy" disease (SCID), *132,* 132–133
Burt, Richard, xiii, 94–98, *95,* 100, 101
Butte, Atul, 228–230

C

C. diff. (*Clostridium difficile*), 113–114
caffeine, and life expectancy, 218
Calbiochem, 231
Calico Labs, 272
California Institute for Regenerative Medicine (CIRM), 41, *41*
California Stem Cell Research and Cures Act (Proposition 71), 41
Calment, Jeanne Louise, 189–190
Cambridge University Hospitals NHS Foundation Trust, 169
Canavan disease, 151
cancer. *see also specific cancers, e.g.:* leukemia
  and alcohol consumption, 218
  James Allison's CTLA-4 research, 72, *73,* 75–76
  CAR-T therapy, 79–83
  CAR-T therapy for ALL, 126–132
  cell senescence in prevention of, 200, 202
  and cellular health, 219–222
  checkpoint inhibitors, 77–79, *78*
  clinical trials for treatments of, 265
  CRISPR-based treatment for, 147, 161
  early immunotherapy studies, 69–72
  as family of related diseases, 71–72
  first use of immunotherapy, 67–69
  hematopoietic stem cells for bone marrow replacement, 10
  immunotherapy for treatment of, 66–84, 120–122, *122*
  malfunctioning genes associated with, 140
  and metformin, 192
  philanthropy supporting research on, 243–245, 273
  pluripotent embryonic stem cells and, 6–7
  RAS proteins in, 234–235
  and regulation of MYC gene expression, 207–208
  stem cell research/treatments, 54–64
  stem cells as drug delivery agents, 61–64
  stem cell treatment for GVHD, 56–60, *58*
  targeting specific cells, 164–169
  telomere lengthening and, 195–197
  treatments for, as antiaging therapies, 203–204
Cancer Moonshot, 227
Cancer Research, 174–175
Cancer Research Institute (CRI), 72
cancer vaccines
  in CAR-T therapy, 125–132
  clinical trials, 83–84
  mechanism of, 125
cardiac fibrosis, 15–16
CardioCell, 18
cardiovascular health, alcohol consumption and, 218
Carnegie, Andrew, 238
CAR-T-19 therapy, 131
Carter, Jimmy, 66–67, *67,* 77
Carter, Rosalynn, 66
CAR-T (chimeric antigen receptor T cell) therapy, 79–83, 126–132, 161
Cas9 enzyme, 142, 144
Catholic Church, xii
CC5 gene, 140
C-Cure, 18
CD (protein), 63
CD3 (protein), 273
CD19 (protein), 80, 81, 127
CD30 (protein), 84
Cedars-Sinai Heart Institute, 16
Cedars-Sinai Medical Center, xiii, 244
celiac disease, 116–118, *117*
cell-based therapies
  for aging, 189, 200
  for allergies, 115
  future of, 136, 276–278
  for gluten intolerance and celiac disease, 118
cell division, *4, 5,* 5–6
cell health, and life expectancy, 219–222, *220*
Cell Reports, 171
cells
  as drugs of the future, 269–281
  as ecosystem, 219–222
cell senescence, *4, 201*
  in the brain, 204–207
  and life expectancy extension, 200–204
cell-surface proteins, in autoimmune diseases, 90–91
Cellular Horizons, xii
cellular medicine
  philanthropy supporting research in, 251
  for treatment of asthma, 109–111
  for treatment of food allergies, 105–109
Celyad, 18

Center for Genetics and Society, 178, 179
Center for International Blood and Marrow Transplant Research (CIBMTR), 57
Center for Pediatric Bioinformatics, 228
Centers for Disease Control and Prevention, 12, 169–170, 172, 266
CETP (cholesterol ester transformer protein) gene, 191
cetuximab (ERBITUX), 167
CFTR gene, 140
Chan, Priscilla, 272
Charpentier, Emmanuelle, 161, 182
checkpoint inhibitors, 77–79, *78*, 125
chemotherapy
  and CAR-T immunotherapy, 126
  healthy cells killed by, 164
  immunotherapy in combination with, 121–122
  in multiple sclerosis treatment, 94, 95
  in myasthenia gravis treatment, 100
  nitrogen mustard as, 262
  stem cells as drug delivery agents, 61–64
Chiari malformation, 138–139
childbirth, death during, 211
Children's Hospital of Philadelphia (CHOP), 126–128
Children's National Medical Center, 139
chimeric antigen receptor T cell therapy. *see* CAR-T therapy
cholesterol ester transformer protein (CETP) gene, 191
cholinesterase inhibitors, 40
CHOP (Children's Hospital of Philadelphia), 126–128
chronic inflammation, 201
chronic myeloid leukemia (CML)
  CRISPR in treatment of, 142–144
  GLEEVEC and, 143, 165–167
Church, George M., 184
CIBMTR (Center for International Blood and Marrow Transplant Research), 57
Cincinnati Children's Hospital, 154
CIRM (California Institute for Regenerative Medicine), 41, *41*
City of Hope (Duarte, California), 62–63
Clara (girl with Leber congenital amaurosis), 138–142, 148
Clinical Microbiology Review, 117–118
clinical trials, 256–268
  and big data, 225
  cancer vaccines, 83–84
  of CAR-T therapy, 131, 132
  of combined therapies, 123–125
  of CRISPR-based gene therapy, 146–148, 159–162
  ethical issues, 264–266
  history, 259–262
  multiple sclerosis treatment, 95–96
  of omalizumab for treatment of allergies, 108–109
  in small patient populations, 266–268
  with targeted treatments, 262–264
clinics, questionable therapies dispensed by, 49–50, 200
Clostridium species, 115
Clostridium difficile (*C. diff.*), 113–114
cloud-based apps, tracking lifestyle with, 234–235
clustered regularly interspaced short palindromic repeats. *see* CRISPR
CML. *see* chronic myeloid leukemia
coffee consumption, and life expectancy, 218
Cohn, Ronald, 159
Cold Spring Harbor Laboratory, 169
Coley, William B., 67–69, *68*
Coley's toxins, 68
colitis, 114
colon cancer
  immunotherapy and chemotherapy for treatment of, 121–122
  oncogene inhibitor for treatment of, 167
colony–collapse disorder, 144–145
colorectal cancer(s)
  drug libraries for treatment of, 231
  oncogene inhibitor for treatment of, 167
Coontz, Sonia Olea, 35–36, 39
cord blood, 10
  Alzheimer's research, 43
  Jenna Gibson and, 257–258
coronary heart disease, 15–18
Cougentakis, Elizabeth, xiii, 98–101, *101*
Creator (film), 185
CRI (Cancer Research Institute), 72
CRISPR (clustered regularly interspaced short palindromic repeats), 118, 140–148, *146*, 163
  Duchenne muscular dystrophy treatment trials, 157–159
  and end of genetic disease, 186
  ethical issues, 176–183
  in human trials, 159–162
  oncogene targeting with, 168–169
  reasons for prohibiting human germ line modification with, 178–180
  staph infection treatment research, 172–175
  Zika treatment with, 171
CRISPR Therapeutics, 161
crizotinib (XALKORI), 167, 231, 264
Cro-Magnon era, life expectancy in, 188
Crow, Sheryl, 54
crowdfunding, 254–255
CTL019 (CAR-T therapy), 127–128
CTLA-4. *see* cytotoxic T-lymphocyte-associated protein 4
CTLA4Ig (drug), 273
cyclosporine, 231
cystic fibrosis
  malfunctioning genes associated with, 140
  ORKAMBI for treatment of, 149
cytotoxic T-lymphocyte-associated protein 4 (CTLA-4), 72, *73*, 75–77, 273

D

Damiano, Edward, 248
Dandy-Walker malformation, 151
dasatinib (SPRYCEL), 203–204
data-sharing collaborations, 226–227
de Grey, Aubrey, 269–272
degrons, 175
dementia, 205. *see also* Alzheimer's disease
  cellular health and, 219

and physical exercise, 214–215
demyelination, in multiple sclerosis, 91–93, *92*
DERL2 gene, 171
"designer babies," 178
"designer humans," 185–186
Diabetes, Obesity and Metabolism, 192
diabetes mellitus, 209. *see also* type 1 diabetes; type 2 diabetes
diet, 217–218, 243–244
disease prevention, big data and, 234–235
DNA. *see also* gene therapy
aging and damage to, 191
cancer due to mutations in, 219–220
and CRISPR, 142, 159
and drug design, 232
editing of, in gene therapy, 141, *141*
in Human Genome Project–Write, 184
mutations and duplication of, 70
nucleotide bases of, 193, 194
telomeres in replication of, *193,* 193–194
of tumor, 226
"Don't Edit the Human Germ Line" (article), 180
dopamine, 50, 52, 229
Doran, Timothy, 144
DOT1L gene, 168
double-blind studies, 108, 261–263
Doudna, Jennifer, 138, 176, 182, 186
Dranoff, Glenn, *82*
Drucker, Peter, 224
drug design, big data and, 232–235
drug development
and big data, 227
for rare diseases, 150, 152
"druggable targets," 276–277
drug libraries, 230–232
drug repositioning, 227–230
drugs, off-label use of, 202–203
Druker, Brian, 165, 166
Dua, Kulwinder S., 28, 30
Duchenne, Guillaume, 154
Duchenne muscular dystrophy, 153–162, *155*
clinical trials in small patient populations, 266–268
CRISPR treatment trials, 157–159
Duke University, xiii, 159
Duke University School of Medicine, 60
dystrophin, 154, 157, 159, 267

E

Eagleman, David, 255
Eales disease, 151
ear, bioprinted, *23*
ecosystem, cells as, 219–222
eczema, 104, *104*
Edison, Thomas, 66
Editas Medicine, 147–148, 160–161
EGFR gene, 227, 230
eggs, allergies to, 106, 107, 144
Einhorn, Thomas, 14–15
Einstein, Albert, 181
Eli Lilly, 203
Ellison, Larry, 272
embryonic stem cells, 6–7
EMC2 gene, 171
EMC3 gene, 171
EML4 gene, 167
Emory-Georgia Tech Healthcare Innovation Program, 249
Emory University, 48, 208
endothelial cells, 203
environmental concerns, 180
epinephrine, 229
epithelial cells, 32
epoetin alfa (PROCRIT), 202–203
ERBITUX (cetuximab), 167
erysipelas, 67
esophageal cancer, 27
esophagus, laboratory-grown, 27–28, *29, 30,* 30–31
Estrada, Stephen, 121–122
estrogen, 264
ethical issues
designer humans, 185–186
with gene therapy, 176–186
and HGP-Write, 183–185
and human germ line, 176–186
with modern clinical trials, 264–266
reasons for prohibiting human germ line modification with CRISPR, 178–180
with stem cells, 3–4
Europe
experimental regenerative medicine in, 11–12
life expectancy in, 188
exercise, and life expectancy extension, 214–216
Expanded Disability Status Scale, 95
Experimental Gerontology, 206, 207
experimental organs, laboratory-grown, 31–34, *33*
experimental studies, clinics involved in, 19
Experiment.com, 254
extracellular matrix, 199
EZH2 gene, 175

F

FA (Fanconi anemia), 104, 151
Facebook, 250, 272
fallopian tubes, 32
Fanconi anemia (FA), 104, 151
FASEB Journal, 196
Fate Therapeutics, 18
5-FC (5-fluorocystosine), 63
FDA approval. *see* US Food and Drug Administration approval
fecal transfer, 114
Federated States of Micronesia, 169
Feinberg School of Medicine, 95
Feldman, Eva, 48, 49
Female Sexual Function Index, 25
Fernandez, Donna, 124–125
Feynman, Richard P., 261
fibrosis, 15–16
Ficken, John, 68
finasteride (PROSCAR, PROPECIA), 203
Fishman, Mark, 275–276
fluorescence in situ hybridization, 103
5-fluorocystosine (5-FC), 63
5-fluorouracil (5-FU), 63, 64
food(s)

genetically-modified, 145, *146*
gluten intolerance/celiac disease and sensitivities to, 116–118, *117, 119*
food allergens, *103*
food allergies
after bone marrow transplants, 102–105
cellular medicine for treatment of, 105–109
CRISPR in prevention of, 144
Treg levels of people with, 109
Food and Drug Administration Safety and Innovation Act, 267, 268
Ford, Henry, 238–239, 252
Fox Chase Cancer Center, 165
Francis, Pope, xii
Fred Hutchinson Cancer Research Center, 56–57
5-FU (5-fluorouracil), 63, 64
Furlong, Christopher, 153–157, 162
Furlong, Pat, *153*, 153–158, 162, 268
Furlong, Patrick, 153–157
Furlong, Tom, 154
fusion mutation, 142
future research and trends, 269–281
bone repair, 279–280
cell-based therapies, 276–278
immunotherapy, 120–136
kidney transplants, 278–279
regenerative medicine, 272–276

G

galactosemia, 151
Gandhi, Mahatma, xv
Gehrig, Lou, 45–46, *46*
Geisel, Theodor Seuss, 237
gene therapy, 138–186
CRISPR and, 140–148
CRISPR for staph infections, 172–175
for Duchenne muscular dystrophy, 153–162
early research, 138–148
ethical issues, 176–186
repairing DNA in rare diseases, 149–162
to stop aging, 207–209
targeting diseased cells, 163–175
Zika vaccine development, 170–171
genetically-modified foods, 145, *146*
germ layers, 6
germ line cells. *see* human germ line
Gersbach, Charles, 159
Gibson, Jenna, 256–258
Gibson, Julie, 256–257
Gillis, Brian, 145
Gilman, Alfred Z., Sr., 262
Gladstone Institutes (San Francisco), 32
Glass, Jonathan, 48, 49, 50
GLEEVEC (imatinib), 143, 165–167, 231
glial cells, *206*, 206–207
glioblastoma multiforme, 61–64
gluten intolerance, 116–118, *117, 119*
Goldstein, Larry, 19, 41–42
Goodman, Louis S., 262
Google, 272
graft-versus-host disease (GVHD), *58*
and bone marrow transplants, 55
stem cell treatment, 56–60, *58*
Great Depression, 239
Groft, Stephen C., 150, 152
Gurdon, Sir John, 7, *8*
gut, microbiome in, 113–114
GVHD. *see* graft-versus-host disease

H

Hailey-Hailey disease, 151
Han, Renzhi, 159
haploinsufficiency, 208
Harris Center for Precision Wellness, 181–182
Harvard Medical School, 184, 248
Harvard University, 182, 230
diabetes type 1 research, 88–89
regenerative medicine research, 274–276
Hawaii, 180
Hayflick limit, 194, 195, 196
HBB gene, 146
HDL cholesterol, 191
health span, extending, 205
heart, laboratory-created tissue for, 32–33
heart disease
dasatinib and quercetin for treatment of, 204
stem cell treatments for, 15–18, *17*
heart transplant, 197–199, *198*
Helmsley, Leona, 247
Helmsley Charitable Trust, 247, 248
Hemacord, xii
hematopoietic stem cells, 9–10, *97*
autologous transplantation of nonmyeloablative, 94–96
resetting the immune system with, 96–98
hemolysin, 172
hemorrhagic stroke, *37, 38*
HER2 (protein), 84
HER2 gene, 263–264
Herceptin (trastuzumab), 84
HgbS gene, 140
HGP-Write (Human Genome Project–Write), 183–185
high-throughput screening, 233
hip pain, 13–15, *14*
hippocampus, 44
Hippocrates, 65, 120
histamines, 144, 229
HIV
adoptive T cell strategies for treatment of, 133
malfunctioning genes associated with, 140
hive-proud honeybees, 144–145
hives, 106, *106*
HLA-DQ proteins, 116
HLA (human leukocyte antigen) proteins, 55, 57
Hodgkin's lymphoma, 84
Home Depot, 249
homeostasis, 210
Homeostatic Capacity Prize, 210
honeybees, CRISPR to prevent colony–collapse disorder with, 144–145
Hospital for Sick Children (Toronto), 159
Hospital for Special Surgery (New York City), xiii, 12–13
Hotta, Akitsu, 159
HRD1 gene, 171

Huang, Junjiu, 176–177
Hughes, Howard, 240
Human Cell Atlas, 227
Human Gene Therapy, 160
human genome
CRISPR in editing of, 146
"language" of, 158
synthesizing, 184
Human Genome Project, 183, 184, 232–233
Human Genome Project–Write (HGP-Write), 183–185
human germ line
designer humans, 185–186
ethical issues with modification of, 176–186
reasons for prohibiting modification with CRISPR, 178–180
human leukocyte antigen (HLA) proteins, 55, 57
Hungerford, David A., 165
Hunter College, 230
huntingtin gene, 140
Huntington's disease, 140
hyaluronic acid, 12
hydrogel capsules, novel beta cells, 89
hydrogels, 280
hygiene hypothesis, 110–111, 115

I

Icahn School of Medicine, 182
Ice Bucket Challenge, 46–47, *47*
IL-6 (interleukin-6), 128
imatinib (GLEEVEC), 143, 165–167, 231
imipramine (Tofranil), 229
immune system, *70*
in autoimmune diseases, 90–91
and cell senescence, 200–201
and GHVD, 57
and multiple sclerosis, 91–94
in treatment of allergies, 105–109
immunoglobulin, 108, 109
immunomodulation, *110*
immunotherapy
allergy treatments, 105–109
for autoimmune diseases, 85–101
cancer treatments, 66–84, 120–122, *122*
cancer vaccines, 83–84
CAR-T therapy, 79–83
checkpoint inhibitors, 77–79, *78*
combined therapies, 123–125
CRISPR technology with, 147
early cancer research, 69–72
first use of, 67–69
future of, 120–136
and hematopoietic stem cells, 96–98
for multiple sclerosis, 91–94, *92*
for SCID, 132–133
for type 1 diabetes, 88–90, 133–136
Imperial College London, 202
induced pluripotent stem cells (iPSCs), xi, 7
heart tissue from, 32–33
kidneys from, 31
Parkinson's research, 52
infants
microbiome of, 114–115
prevalence of genetic disorders in, 149–150
Zika infection in, 170
inflammation, and cancer risk, 221–222
injured tissue, regeneration of. *see* tissue regeneration
innovation, philanthropy and, 237–255
Institute for Aging Research, 190
insulin, 87, 89, 247
Integrated Tissue and Organ Printing system (ITOP), 25–27, *26*
interleukin-6 (IL-6), 128
International Conference on the Progress of Regenerative Medicine, 91
International Space Station, 190
International Stem Cell Corporation, 52
International Summit on Human Gene Editing, 182–183
ipilimumab (YERVOY), *73*, 76–77, 123, 167
iPSCs. *see* induced pluripotent stem cells
Irene (super-ager), 213, 214, 219, 222
Isacson, Ole, 50–52
ischemic stroke, *37, 38*
ISIS-3, 263
ITOP (Integrated Tissue and Organ Printing system), 25–27, *26*
Ivemark syndrome, 151
ixmyelocel-T, 16–17

J

Jagger, Mick, 188
JAMA Neurology, 278
James Buchanan Urological Institute, 238, 244
Jane (super-ager), 213, 214, 219, 222
Japan, 59
Jarcho-Levin syndrome, 151
JCR Pharmaceuticals Co., Ltd., 59
Johns Hopkins Hospital, 139–140, 244
Johns Hopkins University, 238
joint injuries, stem cell treatment for, 10–15
Journal of the National Cancer Institute, 218
June, Carl, 80–83, *81, 82,* 127, 128, 161

K

Kahn, James, 45
KANUMA, 149
KAT2A gene, 168, 169
ketoacidosis, 86
ketones, 86
Key Guardian Award, 242
Key Philanthropy Award, 242
KEYTRUDA (pembrolizumab), 67, 77, 79
kidney, 21, 22, 31–32
kidney cancer, 230
kidney transplants, future of, 278–279
kinase map (K-MAP), 231–232
King, Stephen, 184
Klotho enzyme, 209
K-MAP (kinase map), 231–232
knees, 11–13, *13*
Kurtzberg, Joanne, xiii, 60
Kyoto University, 159

L

lactobacilli, 117, 118

The Lancet, 24–25
Langone Medical Center (New York University), 13–15
Lasker Awards, 72, *74,* 75
latex allergy, 104
Leber congenital amaurosis (LCA)
characteristics of, 139–140
CRISPR trials, 147–148, 159–161
viral-based gene therapy for, 141
Leisure World, 213
Leisure World Cohort Study, 214
leukemia, 54–60
allergies in patients after treatment for, 102–104
and bone marrow transplants (*see* bone marrow transplants)
CAR-T therapy for ALL, 126–132
CAR-T therapy for chronic lymphocytic leukemia, 80–83
Jenna Gibson and, 256–258
hematopoietic stem cells for bone marrow replacement, 10
origins of term, 80
Levi-Montalcini, Rita, 276
Levine, Bruce, *82*
life expectancy extension, 188–211, 211–222
20th-century scientific advances, 211–212
accelerating innovation process with funding, 209–210
cell senescence and, 200–204
cellular health and, 219–222, *220*
future issues, 269–272
gene editing to stop aging, 207–209
lifestyle changes for, 214–219
metformin, 191–193
90-Plus Study, 213–218
questionable therapies, 200
with replacement organs, 197–199
"super-agers," 189–191
targeting cell senescence in the brain, 204–207
telomere repair, 193–197
lifestyle changes
big data and, 234–235
for life expectancy extension, 212, 214–219
Lind, James, 259–260, 265
Little, Arthur D., 2
liver, GVHD and, 58
liver organoid, *33*
Longevity Demonstration Prize, 209–210
Longevity Genes Project, 190
Lou Gehrig's disease. *see* amyotrophic lateral sclerosis (ALS)
Lower, Richard, 197
LSD, 255
Lucile Packard Children's Hospital, 228
Ludwig, Bill, 80–83, *82*
lung cancer
clinical trials in small patient populations, 267
combined therapy for, 124–125
GLEEVEC in treatment of, 166–167
non-small-cell, 227–230, *229*
oncogene inhibitor for treatment of, 167
lungs, modeling of, with regenerative cellular techniques, 33, 34
lymphoma, 10
LYNPARZA (olaparib), 167
lysosomal acid lipase deficiency, 149

M

MABs (monoclonal antibodies), 84
malnourishment, 211
Manning, Peyton, 10–11
MAPK protein, 231
Marban, Eduardo, xiii, 16
Marcus, Bernie, 248–250, *249*
Marcus Foundation, 249–250
Marcus Institute for Brain Health, 250
Margolis, David, 133
Martin, Case, *132,* 133
Massachusetts Institute of Technology (MIT), 269
Massachusetts Institute of Technology (MIT) Broad Institute, 182
Massella, Luke, 20–22
Max Planck Institute, 172
Max Planck Institute for Infection Biology (Berlin), 32, 182
Mayer-Rokitansky-Küster-Hauser Syndrome (MRKHS), 24–25
Mayo Clinic, 30, 201, 204
McCarthy, Joe, 45
McLean Hospital Neuroregeneration Research Institute, 50–51
McMaster University, Farncombe Family Digestive Health Research Institute, 116
MCP-150-IM cell therapy, *17,* 17–18
M.D. Anderson Cancer Center, 72
Medical College of Virginia, 197
Medical College of Wisconsin, 28
medical research, big data and, 224–236
Medicare, 205
Meihaus, Grace, xiii, 96–98, *99,* 101
melanoma
Sharon Belvin and, 75–77
combined therapy for treatment of, 123–124
oncogene inhibitor for treatment of, 167
Melton, Douglas A., 85–90, *90*
Melton, Gail, 85, 86
Melton, Sam, 85–87
Memorial Hospital (New York), 67–69
Memorial Sloan Kettering Cancer Center, 123, 132, 268
MEN1 gene, 168
mental illness, philanthropy supporting research on, 250
mesenchymal stem cells, *8,* 9
Alzheimer's research, 43–45
glioblastoma treatment, 64
GVHD treatment, 59–60
and ixmyelocel-T, 16
MCP-150-IM cell therapy, *17,* 17–18
stroke treatment, 37, 38
Mesoblast, *17,* 17–18, 59, 60
messenger RNA (mRNA), 173–174, 196
metformin, 191–193, 230
Methuselah gene, 191
mice
3D-printed organ tests in, 26

Alzheimer's research, 42, 43
antiaging therapy research in, 203–204
antigen-specific Treg therapy in, 111
bone repair research, 280
cancer risk for, 219, 221
cell senescence research in, 201–202
CRISPR and microRNA in, 174
Duchenne muscular dystrophy research, 159
genetic research on CML in, 142–144
MYC gene research in, 208
NMN research in, 209
microbiome, *113*
allergies and, 112–115
antibiotics and, 164
celiac disease and, 116–118
microRNA, 173–174
Milken, Michael, and prostate cancer research, 243–245
Million Woman Study, 218
"mini-transplant," stem cell, 98
MIT (Massachusetts Institute of Technology), 269
MIT (Massachusetts Institute of Technology) Broad Institute, 182
mitochondria, 270
MMAE, 84
Molecular Therapy, 160
mongoose, 180
monoclonal antibodies (MABs), 84
mood disorders, 114
Motherboard, 271
Mt. Sinai Hospital, 182
Mozambique, 212
"Mr. SR," 57–58
MRKHS (Mayer-Rokitansky-Küster-Hauser Syndrome), 24–25
mRNA (messenger RNA), 173–174, 196
MS. *see* multiple sclerosis
MSC-100-IV, 59, 60
mTOR gene, 209
multiple sclerosis (MS), 91–94, *92, 94*
demyelination in, 91–93, *92*
philanthropy supporting research on, 250
prevalence of, 93, *94*
stem cell treatment research, 94–96
Murdoch Childrens Research Institute (Brisbane, Australia), 31–32
muscle injuries, stem cell treatment for, 10–15
muscular dystrophy, *156. see also* Duchenne muscular dystrophy
mustard gas, 262
mutations, 70
myasthenia gravis, xiii, 91, 98–101
MYC gene, 207–208
myelin, 91–93

N

Nagler, Cathryn, 115
Nair, Girish, 52
Napster, 250, 272
National Cancer Database (NCDB), 225–226
National Cancer Institute, 239
National Heart Institute, 239
National Institute of Health, 239
National Institute of Mental Health, 239
National Institutes of Health (NIH), 90, 152, 208–209, 232, 240, 241, 247, 249, 253–255
National Organization for Rare Disorders (NORD), 151
natural selection view of cancer, 221
Nature (journal), 171, 177, 180, 202, 217, 234
Nature Biotechnology, 26–27, 169
Nauts, Helen Coley, 72
NCDB (National Cancer Database), 225–226
Neal, Joel, 230
Nebuchadnezzar, 259
neck injuries, 10–11
NEK1 gene, 47
nerve growth factor (NGF), 276–278
neurogenesis, 44
neurons
demyelination of, *91,* 91–92
senescence for, 206
Neuroregeneration Research Institute, 50–51
neurotransmitters, 229
New England Journal of Medicine, 169, 248
Newsweek, 271
New York University, 13–15
NGF (nerve growth factor), 276–278
nicotinamide mononucleotide (NMN), 209
NIH. *see* National Institutes of Health
90-Plus Study, 213–218, 222
nitrogen mustard, 262
nivolumab (OPDIVO), 123–125, 167
Nixon, Richard M., *71*
NMN (nicotinamide mononucleotide), 209
Nobel Prize
Sir John Gurdon, 7
Rita Levi-Montalcini, 276
E. Donnall Thomas, 54–55, *56*
Shinya Yamanaka, 7
"noise," genetic, 228
non-Hodgkin's lymphoma, 57–58, 262
non-small-cell lung cancer (NSCLC), 227–230, *229*
NORD (National Organization for Rare Disorders), 151
norepinephrine, 229
Northwestern University, xiii, 95, 97, 100
Novartis, 175
Nowell, Peter C., 142–143, 165
NSCLC (non-small-cell lung cancer), 227–230, *229*
NTRK gene, 227
nucleotide bases, DNA, 193, 194
NULOJIX (belatacept), 273
NurOwn therapy, 48
nut allergies, 107
nystagmus, 138, 139

O

Obama, Barack, 227
obese people, life expectancy of, 216
off-label use, of drugs, 202–203
off-target effects, of CRISPR, 145–147
Ohio State University, 159
olaparib (LYNPARZA), 167
Olson, Eric N., 158–159
omalizumab (XOLAIR), 108–109

oncogenes, 166
Oncology, 57–58
Oncology Research Information Exchange Network (ORIEN), 226–227, 267
OPDIVO (nivolumab), 123–125, 167
Oracle, 272
Oregon Health & Science University, 165
ORENCIA (abatacept), 273
organoids, 32, *33*
Organovo, 31–32
organs, laboratory-grown, 20–34
  bladder, 21–22, *22*
  esophagus, 27–28, *29, 30,* 30–31
  experimental models, 31–34, *33*
  life expectancy extension with, 199
  printing body parts, 25–27, *26*
  salamander research, 23
  vaginal organs, 24–25
organ transplants, *198*
  life expectancy extension with, 197–199
  philanthropy supporting research on, 273
ORIEN (Oncology Research Information Exchange Network), 226–227, 267
ORKAMBI, 149
Ornish, Dean, 194
orphan diseases, 150. *see also* rare diseases
orphan drugs, 150, 152
osteoarthritis, stem cell treatment study, 12–13
osteonecrosis, 14
O'Toole, Peter, 185
ovarian cancer, oncogene inhibitor for treatment of, 167
overweight people, life expectancy of, 216

P

p53 gene, 174
PAI-1 (plasminogen activator inhibitor-1), 197
Palo Alto Investors, 209
Palo Alto Longevity Prize, 209–210
Panzirer, David, 245–248, 253
Panzirer, Morgan, 245–248, *246*
PAP (prostatic acid phosphatase), 125
Pardoll, Drew, 120
Parent Project Muscular Dystrophy, 157, 268
Parker, Sean, 147, 250–251, *251,* 272
Parker Foundation, 251
Parker Institute for Cancer Immunotherapy, 272
Parkinson, James, 50
Parkinson's disease, xv, xvi, 50–52, 250
Patel, Amit N., 16–17
patients, segmenting of, in clinical trials, 265–266
PCF (Prostate Cancer Foundation), 244–245
PCF Scientific Retreat, 245
PD-1 protein, in CAR-T therapy, 131–132
PD-1 receptor blockers, 78, 79
PD-L1 blockers, 121
Peale, Robert C., 102
peanut allergy, 104, 105, 115
Pearce, David, 133, *134,* 135–136
pembrolizumab (KEYTRUDA), 67, 77, 79
pentamidine, 230
Peto's paradox, 219, 221
PetriDish, 254
Pet Sematary (King), 184
PF-02341066 (drug), 167
Pfizer, 203
PGC-1a gene, 208
phase I trials, of Treg therapy, 134
phase II trials
  of NGF, 278
  of Treg therapy, 134–135
Philadelphia chromosome, 142, 143, 165, *166*
philanthropy, 237–255
  anti-aging research, 209–210
  crowdfunding, 254–255
  Bernie Marcus, 248–250, *249*
  medical success stories, 245–248
  by megaphilanthropists, 272–274
  Michael Milken and prostate cancer research, 243–245
  Sean Parker, 250–251, *251*
  pilot funding, 253–254
  T. Denny Sanford, *252,* 252–253
  seed grants, 253–254
PI3Ka gene, 175
pilot funding, 253–254
placebos, 261
plant-based foods, and life expectancy, 218
plasminogen activator inhibitor-1 (PAI-1), 197
platelet-rich plasma (PRP), 12
PLCO Cancer Screening Trial, 263
pluripotent embryonic stem cells, 6–7
pluripotent stem cells, 42
portion size, and life expectancy, 218
Precision Medicine Initiative, 227
printing of body parts, 25–27, *26*
probiotics, 115
Proceedings of the National Academy of Sciences, 221
PROCRIT (epoetin alfa), 202–203
progesterone, 263
proinflammatory cytokines, 201
Project MinE, 46
PROPECIA (finasteride, PROSCAR), 203
Proposition 71 (California Stem Cell Research and Cures Act), 41
prostate cancer, 243–245, 264
Prostate Cancer Foundation (PCF), 244–245
prostate disease, 238
prostatic acid phosphatase (PAP), 125
PROSTVAC (cancer vaccine), 84
Protein & Cell, 146, 177
proteins
  as allergens, 105–106
  binding pocket in, 169
PROVENGE (sipuleucel-T), 83, 125
PRP (platelet-rich plasma), 12
Pryor, Richard, *93*
PSA test, 243
Pseudomonas aeruginosa, 116–117
Public Health Service, 262

Q

Queen Mary University of London, 209
quercetin, 203–204

R

rabbits, 180

radiation treatment, with bone marrow transplants, 55
randomized-control trials, 260–263
Ransdell Act, 239
rare diseases
CRISPR for treatment of, 159–162
drug development for, 150, 152
Duchenne muscular dystrophy as, 153–159
gene therapies for, 149–162
genetic disorders as, 149–150
RAS protein, 234–235
rats
3D-printed organ tests in, 26–27
whole-organ decellularization in, 199
Ravasi, Gianfranco, *242*
recessive autosomal genetic disorders, 139
red meat consumption, 218
regeneration of tissue. *see* tissue regeneration
regenerative medicine, 4–5
future of, 272–276
"spare parts" approach to, 275
regulatory T cells (Tregs), *110,* 273
asthma treatment with, 109–111
type 1 diabetes treatment with, 133–136
relapsing-remitting MS, 92, 95
repositioning (drug repositioning), 227–230
research
big data and, 224–236
clinical trials, 256–268
philanthropy and, 237–255
resveratrol, 209, 218
Reuter, Gary, 144–145
Rheingold, Susan R., *126,* 126–129, 131
rhesus monkeys, 169
rheumatoid arthritis, 91, 128
Richards, Keith, 188
risks, with CRISPR gene editing, 145–147
rituximab (RITUXAN), 84
RNA, 62, 142, 159
Rockefeller, David, *239*
Rockefeller, John D., Sr., 238, *239*
ROGAINE, 230
ROS1 gene, 227
ROS1+ lung cancer, 267
Ross, Steve, 243
Royal Melbourne Hospital (Australia), 52
RPE65 gene, 139–140
Rutgers University, 248

S

Safe Kids Worldwide, 12
salamanders, 23
sample size, statistics and, 224–226
Sanfilippo, Fred, 249, 250
Sanford, T. Denny, 133, *252,* 252–253
Sanford Center, *135*
Sanford Health, 252–253
Sanford Project, 7, 134–136, 253
Sanford Research, 133
SCID (severe combined immunodeficiency), *132,* 132–133
Science (journal), 177, 183, 211
Science and Business of Regenerative Medicine symposium, 274
Scientific Reports, 172, 174
scurvy, 259–260, 265
Sean N. Parker Autoimmune Research Lab, 251
Sean N. Parker Center for Allergy & Asthma Research, 251
Seattle Children's Hospital, 256
Seattle Public Health Service Hospital, 55
secondary progressive MS, 92
seed grants, 253–254
SEL1L gene, 171
selection bias, 260
Selleck Chemicals, 231
senescence. *see* cell senescence
senolytic drugs
Alzheimer's disease treatment with, 204–205, 207
repurposing other drugs as, 202–204
SENS Research Foundation, 271
serotonin, 229
Seuss, Dr., 237
Seventh-day Adventists, 2112
severe combined immunodeficiency (SCID), *132,* 132–133
SGK1 gene, 230
shield compounds, 175
signaling networks, 203
Silicon Valley, medical philanthropy in, 250
Simpson, Ericka, 49–50
sipuleucel-T (PROVENGE), 83, 125
sirtuins, 209
Skerrett, Donna, 17–18, 59
skin testing, for allergies, *107*
skipped team references with sports stars, 10–12
Sleeper (film), 185
smoking, 211, 213
social engagement, and life expectancy, 216–217
sodium trideceth sulfate, 199
"A Sound of Thunder" (Bradbury), 180–181
"spare parts" approach to regenerative medicine, 275
SPCS genes, 171
spina bifida, 20–22
Spivak, Marla, 144–145
SPRYCEL (dasatinib), 203–204
Spurgeon, Charles, 85
"squaring of the life span," 216
Stahl, Ruth, 213, 219, 222
Stanford University, 108, 196, 228, 230, 231, 251
Stanford University School of Medicine, 37–39
staphylococcus (staph) infections, 172–175
statistics, and sample size, 224–226
Stein, Fred, 67
Steinberg, Gary, 38–39
stem cells, *9*
ALS research, 47–50
Alzheimer's treatment, 40–45, *42, 44*
bone repair treatment, 279–280
for brain/nervous system disorders, 35–53
cancer research/treatments, 54–64
for chronic muscle/joint injuries, 10–15
clinics offering treatment, 19
as drug delivery agents, 61–64
Duchenne muscular dystrophy treatment, 158–159
embryonic, 6–7
GVHD treatment, 56–60

for heart repair, 15–18
iPSCs, 7
mesenchymal, *8*
"mini-transplant" of, 98
multiple sclerosis treatment research, 94–96
Parkinson's treatment, 50–52
and regenerative medicine, 4–5
resetting the immune system with, 96–98
for stroke treatment, 37–39
structure/function of, 5–10
and tissue regeneration, 3–18
for type 1 diabetes treatment, 88–90
in whole-organ decellularization, 199
stem cell treatment study, 12–13
STI571 (compound), 165–166
Stock, Gregory, 181–182
streptomycin, 260–261
Strickland, Sabrina, xiii, 13
stroke, 35–39, *37*, *38*
suicide, by MS patients, 93
"suicide gene," 64
Summers, Lawrence H., 274–275
"super-agers," 189–191, 214
surrogate markers, 267
syphilis, 262
systemic scleroderma, xiii, 96–98

T

targeted treatments, clinical trials with, 262–264. *see also* gene therapy
Targeting Aging with Metformin (TAME) trial, 191–192
Taylor, Doris, 199
T cells, 72, *73*
allergies in patient after treatment for diseases related to, 104, 105
bispecific T-cell engagers, 125
and CAR-T therapy, 80–81
in CRISPR-based cancer treatment, 147
and PD-1 receptor blockers, *78*, 79
provoking, in treatment of asthma, 110–111
TECENTRIQ (atezolizumab), 121–122
Technology Review, 269
T effector cells, 109, 134, 135
telomerase, 194–196
telomeres, *193*, 193–197, *195*
TEMCELL® HS Inj., 59
temozolomide, 61
TERT gene, 196, 208
testosterone, 263–264
Texas Children's Hospital, 133
Texas Heart Institute, 199
Thiel, Peter, 272
Thomas, E. Donnall, 54–57, *56*
3D printing, xv, 25–27, *26*
tissue regeneration, 2–18
heart repair, 15–18
muscle/joint injuries, 10–15
TKIs. *see* tyrosine kinase inhibitors
tocilizumab (ACTEMRA), 128
Tofranil (imipramine), 229
Tolstoy, Leo, 189
toxins, accumulation of, 270
transcription factors, 207
Translational Neuroscience Institute, 277
transplantation, autologous nonmyeloablative hematopoietic stem cell, 94–96
trastuzumab (Herceptin), 84
traumatic brain injury, philanthropy supporting research on, 250
Treatise of the Scurvy (Lind), 259–260
Tregs. *see* regulatory T cells
T-Rex study, 134–136
tumors, 226
Tuskegee Experiment, 262
Tuskegee Institute, 262
Tuszynski, Mark, 277–278
type 1 diabetes, 86–91, *88*
Morgan Panzirer and, 245–248
philanthropy supporting research on, 253, 273
Sanford Project, 7
stem cell treatment research, 88–90
Treg therapy for, 111, 133–136
type 2 diabetes
metformin in treatment of, 191, 192
Morgan Panzirer and, 246
tyrosine kinase inhibitors (TKIs), 165–166
drug design, 233–234
drug libraries, 230–232

U

UBE2G2 gene, 171
UBE2J1 gene, 171
umbilical cord blood. *see* cord blood
United Kingdom, antibiotics in, 261
United Network for Organ Sharing, 279
United States
life expectancy in, 188–189, 211–212
stem cell research in, xi–xii
Unity Biotechnology, 204
University of California, Berkeley, 32, 72, 182
University of California, Irvine, 213, 214
University of California, San Diego, 41, 277, 278
University of California, San Francisco, 134, 251, 279
University of Chicago, 165
University of Colorado, 234
University of Colorado Anschutz Medical Campus, 250
University of Colorado Cancer Center, 231
University of Michigan, 33, 34, 48
University of Minnesota, 197
University of Munich, 102
University of North Carolina, 133
University of Pennsylvania, 147, 161, 165
University of Pennsylvania Abramson Cancer Center, 80–83
University of Pittsburgh, 37
University of Rochester, 280
University of Texas Southwestern, 158–159
University of Utah, 16–17
University of Virginia, 234
US Census Bureau, 212
US Food and Drug Administration (FDA)
approval, 11
for belatacept, 273
for cell-based treatments, xii, 60, 101

clinical trials for, 30, 31, 244–245
for experimental procedures, 28
and Food and Drug Administration Safety and Innovation Act, 267
for GLEEVEC, 166
for imipramine, 229
for leukemia treatments, 257
for metformin, 191, 192
for nitrogen mustard, 263
and off-label uses of drugs, 202
and philanthropy-academic partnerships, 253
for treatments of rare diseases, 150, 152, 162

V

vaccines, anti-cancer, 83–84, 125–132
vaginal organs, laboratory-grown, 24–25
Vanderbilt University Medical Center, 279
van Deursen, Jan, 201–202
van Gogh, Vincent, 189
Vassiliou, George, 169
Vatican, 242, *242*
vemurafenib (ZELBORAF), 167
Verdú, Elena, 117
Vericel Corporation, 16, 17
Vetter, David, *132*
VIAGRA, 230
villi, intestinal, 116
virtual drug docking, 234–235
viruses, in gene therapy, 140–141
vitamin C, 265
vorinostat (ZOLINZA), 133

W

Wake Forest Institute for Regenerative Medicine, 23
Washington Post, 278
Washington University School of Medicine, 170–171
Washkansy, Louis, 198
Watson, James, 137
weapon, gene editing as potential, 181–182
weight, and life expectancy, 216
Wellcome Trust Sanger Institute, 168
Whitehead, Emily, 126–128
Whitehead, Kari, 126
Whitehead, Tom, 126
whole-organ decellularization, 199
Wilkins, Brittany, 131
Wilkins, Nicholas, 129–131, *130*
Williams, Mary Elizabeth, 122–124
Williams, Ted, 224
Withers, Dominic, 202
Wolchok, Jedd, *76,* 76–77
The World Factbook, 211–212
World Medical Association, 262
World War I, 190, 262
World War II, 190
Wright, Bob, 250
Wright, Suzanne, 250

X

XALKORI (crizotinib), 167, 231, 264
X chromosome, 154–155
XOLAIR (omalizumab), 108–109
Xp21 gene, 157–158

Y

Yale University, 134, 234
Yamanaka, Shinya, 7
Yap (island), 169, 170
YERVOY (ipilimumab), 73, 76–77, 123, 167
Young, Hugh Hampton, 238
Yun, Joon, 209–210

Z

ZELBORAF (vemurafenib), 167
Zhang, Feng, 147, 159–161, 182
Zika virus, 169–171, *170*
zinc finger nucleases, 141
ZOLINZA (vorinostat), 133
Zuckerberg, Mark, 272
zygote, 5–6

ABOUT THE AUTHORS

ROBIN L. SMITH, MD is a global thought leader in regenerative medicine, one of the fastest growing segments of modern-day medicine. She received her MD from the Yale School of Medicine and an MBA from the Wharton School of Business. During her tenure as CEO of the NeoStem family of companies (NASDAQ: NBS), which she led from 2006 to 2015, she pioneered the company's innovative business model, combining proprietary cell therapy development with a successful contract development and manufacturing organization. Dr. Smith raised over $200 million, completing six acquisitions and one divestiture while the company won an array of industry awards and business recognition, including a first-place ranking in the tri-state area (two years in a row) and eleventh place nationally on Deloitte's Technology Fast 500 and Frost & Sullivan's North American Cell Therapeutics Technology Innovation Leadership Award.

In 2008 Dr. Smith founded the Stem for Life Foundation (SFLF), a nonpartisan 501(c)3 educational organization devoted to fostering global awareness of the potential for regenerative medicine to treat and cure a range of deadly diseases and debilitating medical conditions, as opposed to merely treating their symptoms. In 2010, in order to bring the charity's mission to a global audience, Dr. Smith forged a historic, first-of-its-kind partnership with the Vatican. As part of this relationship, the Vatican and SFLF collaborate to create high-profile initiatives that help catalyze interest and development of cellular therapies that could ultimately reduce human suffering on a global scale.

Dr. Smith maintains a regular column on these topics for the *Huffington Post* and is coauthor of *The Healing Cell: How the Greatest Revolution in Medical History Is Changing Your Life*. She is a winner of the 2014 Brava! Award, which recognizes top women business leaders in the Greater New York area. She was also a finalist for the 2014 EY Entrepreneur of the Year Award for the New York area, recognizing entrepreneurs who demonstrate excellence and success in the areas of innovation, financial performance, and personal commitment to their businesses and communities. In April 2016, Pope Francis awarded Dr. Smith Dame Commander with Star Pontifical Equestrian Order of Saint Sylvester Pope and Martyr. Dr. Smith was awarded the Lifetime Achievement in Healthcare and Science Award by The National Museum of Catholic Art & Library in May 2017.

Dr. Smith was appointed as clinical associate professor, Department of Medicine at the Rutgers, New Jersey Medical School in 2017. In addition, Dr. Smith has extensive experience serving in executive and board-level capacities for various medical enterprises and health care–based entities. She currently is chairman of the board of directors of MYnd Analytics (NASDAQ: MYND, formerly CNS Response); serves on the board of directors of Rockwell Medical (NASDAQ: RMTI), and Prolung DX; the advisory board of Hooper Holmes (OTCQX: HPHW); and is cochairman of the Life Sci advisory board on gender diversity. She is vice president and member of the board of directors of the Science and Faith STOQ Foundation in Rome and serves on Sanford Health's International Board and the board of overseers at the NYU Langone Medical Center in NYC. She previously served on the board of trustees of the NYU Langone Medical Center and is a past chairman of the board of directors for the New York University Hospital for Joint Diseases and was on the board of directors of Signal Genetics (NASDAQ: SGNL) and BioXcel Corporation.

As a business leader, entrepreneur, doctor, and philanthropist, Dr. Smith is uniquely positioned to lead the global health care industry into the cellular future—a future where the cells of our bodies will stand as the foundation for a wide array of cures, and big data assists physicians to determine the best course of treatment for individual patients.

© Gillian Fry/Guerrero Howe

MAX GOMEZ, PHD, one of TV's most respected medical journalists, has produced award-winning health and science segments for network stations in New York and Philadelphia. Dr. Max has reported for *Dateline*, the *Today Show*, and *48 Hours*. Over more than three decades, he's earned nine Emmy Awards, three NY State Broadcaster's Association Awards, and UPI's Best Documentary Award.

In addition to New York City's Excellence in Time of Crisis award for his September 11 coverage, Dr. Max has been singled out nationally for special award recognition by the Leukemia and Lymphoma Society and the National Marfan Foundation. He was also named the American Health Foundation's Man of the Year.

A noted moderator and speaker, Dr. Max has been the regular moderator for Memorial Sloan Kettering's Cancer Smart public education series. He was also an organizer and moderator for two international conferences on adult stem cells hosted by and held at the Vatican. Dr. Max also trains physicians nationally in public speaking and presentation, speaking to groups as diverse as the National Cancer Institute and the College of American Pathologists.

Dr. Max is the coauthor of *The Healing Cell: How the Greatest Revolution in Medical History Is Changing Your Life*, a layman's guide to the medicine of the future, showcasing a wide array of emerging adult stem cell breakthroughs, including their ability to repair damaged hearts and organs, restore sight, kill cancer, cure diabetes, heal burns, and stop the march of degenerative diseases, such as Alzheimer's, multiple sclerosis, and Lou Gehrig's disease. Covering stem cell applications in everything from orthopedics to heart disease, cancer, and HIV, it includes a special address by Pope Benedict XVI urging increased support and awareness for advancements in adult stem cell research in order to alleviate human suffering.

Dr. Max is also the coauthor of *The Prostate Health Program: A Guide to Preventing and Controlling Prostate Cancer*. The 384-page book explains how an innovative program consisting of diet, exercise, and lifestyle changes may help prevent prostate cancer.

Among the boards he serves or has served on are:

- The Stem for Life Foundation
- American Heart Association, national and regional boards
- Past chair of the national communications committee for the American Heart Association
- Past vice chair of *Princeton Alumni Weekly*'s board
- Science Writers Fellowship at the Marine Biological Laboratory in Woods Hole, MA
- Crohn's and Colitis Foundation of America—Long Island Chapter
- The Partnership for After School Education, a citywide group of 1,600 community-based programs that tutor and mentor students

- Past chair of the communications committee for the American Association for the Advancement of Science
- Cognitive Warriors, a Cura Foundation

Dr. Max also mentors undergraduate journalism students, as well as medical students and physicians who are interested in medical journalism. He is an honors graduate of Princeton University and the Wake Forest University School of Medicine (PhD in Neuroscience) and was an NIH postdoctoral fellow at Rockefeller University.